MAKING
MATHEMATICS
PRACTICAL

An Approach to Problem Solving

MAKING MATHEMATICS PRACTICAL

An Approach to Problem Solving

Toh Tin Lam • Quek Khiok Seng
Leong Yew Hoong • Dindyal Jaguthsing • Tay Eng Guan

Nanyang Technological University, Singapore

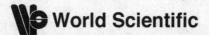

NEW JERSEY • LONDON • SINGAPORE • BEIJING • SHANGHAI • HONG KONG • TAIPEI • CHENNAI

Published by

World Scientific Publishing Co. Pte. Ltd.

5 Toh Tuck Link, Singapore 596224

USA office: 27 Warren Street, Suite 401-402, Hackensack, NJ 07601

UK office: 57 Shelton Street, Covent Garden, London WC2H 9HE

British Library Cataloguing-in-Publication Data
A catalogue record for this book is available from the British Library.

ISBN-13 978-981-4355-00-1
ISBN-10 981-4355-00-3

Printed in Singapore.

Preface

Alan Schoenfeld wrote in the 2007 special issue on problem solving of the journal ZDM that the current focus should lie in translating decades of theory building about problem solving into workable practices in the classrooms:

> That body of research—for details and summary, see Lester (1994) and Schoenfeld (1985, 1992)—was robust and has stood the test of time. It represented significant progress on issues of problem solving, but it also left some very important issues unresolved. ... The theory had been worked out; all that needed to be done was the (hard and unglamorous) work of following through in practical terms.

This book represents an attempt at making problem solving practical in the mathematics classroom. It is the result of "hard and unglamorous" work in a few Singapore secondary schools. It sets out within these pages an approach to the teaching and learning of mathematical problem solving at high school level. The model of mathematical problem solving that underpins this approach is drawn mainly from the works of George Pólya and Alan Schoenfeld, but which we have tempered with suggestions based on our experience in teaching problem solving to Singapore students.

As the science practical lesson is very much a mainstay in Science education, we intend to use mathematics 'practical work'—going through the process of Pólya's stages using the worksheet—to trigger a paradigmatic shift in raising the status of problem solving as integral to mathematical learning. A feature of the "practical" approach is the use of a practical worksheet to guide the problem solving process. The worksheet focuses the solver's attention on the key states in problem solving. For example, the solver needs to control or manage the solving process and, to do so, may have to address not only the subject matter knowledge but also his or her frustration in not being able to reach a solution. Another important part of the practical worksheet, Check and Expand, encourages the student to work mathematically by checking if the answer reached is reasonable and by generating other mathematical problems based on the one just solved.

To acknowledge that assessment drives the curriculum is an important step to implementing valued attributes into the curriculum. Problem solving is valuable. Process skills are valuable. Looking back at a problem is valuable. To help students value these attributes, we need to show explicitly that they are indeed valued, and this, at least for the students, means that they must be assessed and graded. To this end, an assessment rubric has been carefully designed as an indispensable companion to the practical worksheet.

We hope you will find this book useful in your journey into the exciting world of teaching and learning mathematical problem solving.

Toh Tin Lam, Quek Khiok Seng, Leong Yew Hoong,
Jaguthsing Dindyal, Tay Eng Guan
January 2011

Contents

Chapter 1

Mathematical Problem Solving

1.1 Introduction

Mathematical problem solving has been established as the central theme of the Singapore primary and secondary mathematics curriculum since the 1990s. The primary aim of the curriculum is to develop students' ability to solve mathematics problems.

Curriculum documents state that problem solving heuristics are to be taught in our primary and secondary school mathematics classes. Singapore teachers have been provided with preservice preparation or professional development in teaching problem solving, drawing on overseas resources as well as those developed locally. However, these resources tend to emphasise the learning of heuristics but do not focus on mathematics content at a deep level or the kind of mathematical thinking used by mathematicians such as conjecturing and proving.

Towards upper secondary and junior college levels, on the other hand, students tend to concentrate on national exam-type mathematics problems, and in our experience, the heuristics learned at lower levels tend to be ignored by both students and teachers instead of being applied in their mathematical engagements.

To return problem solving to its focal role in the curriculum, we need to build teachers' capacity for problem solving and its teaching, and to develop materials and teaching resources for the explicit teaching of problem solving in the environment of deep mathematics content at the higher school levels. This booklet is a contribution towards this goal. It presents a complete basic problem solving module which can be implemented in a high school curriculum.

The module materials consist of the following:
- Ten one-hour lesson plans
- Templates for a practical worksheet and an assessment rubric
- A collection of problems with
  - accompanying complete solution(s),
  - heuristics to be highlighted,
  - Pólya's stages to be highlighted,
  - suitable scaffolding that a teacher can give to the student,
  - suggested adaptations, extensions and generalizations to the problem,
  - anticipated student responses, and
  - notes on assessment of and for problem solving.

As a necessary preamble to the module materials, we shall first introduce Pólya's model of problem solving via a worked example in Section 1.2. In Section 1.3, we then gird this model with the framework of Schoenfeld which helps to bring into focus the different aspects that contribute to success in problem solving. Finally, in Section 1.4, we explain our approach to teaching problem solving via a 'practical' paradigm.

1.2 Pólya's Model of Problem Solving

We would suggest that any attempt at mathematical problem solving requires a model to which the problem solver can refer, especially when he or she is unable to progress satisfactorily. Good problem solvers have intuitively built up their own problem solving models.

However, for two reasons, a problem solving model that is made explicit would be immensely helpful. Firstly, the typical problem solver who has no model may find that a model helps in regulating his problem solving attempt. Secondly, even a good problem solver will find the structured approach of a model useful and an earlier introduction to the process will help him progress faster in his mathematical development. Consider what Alan Schoenfeld, a mathematician and mathematics educator has to say in the preface to his book *Mathematical Problem Solving*:

> In the fall of 1974 I ran across George Pólya's little volume, *How to Solve It*. I was a practising mathematician, a few years out of graduate school and happily producing theorems in topology and measure theory. Pólya wrote about problem solving, more specifically about the strategies used by mathematicians to solve problems ... My first reaction to the book was sheer pleasure. If, after all, I had discovered for myself the problem-solving strategies described by an eminent mathematician, then I must be an honest-to-goodness mathematician myself! After a while, however, the pleasure gave way to annoyance. These kinds of strategies had not been mentioned at any time during my academic career. Why wasn't I given the book when I was a freshman, to save me the trouble of discovering the strategies on my own?

We shall now describe the essential features of the problem solving model proposed by George Pólya, an eminent Hungarian mathematician. The process can be summarized in the diagram below.

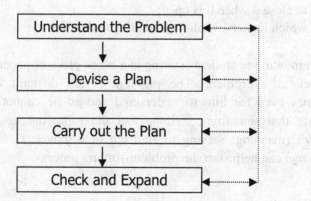

The model resembles a flowchart with four components, *Understand the Problem*, *Devise a Plan*, *Carry out the Plan*, and *Check and Expand*, which are hierarchical but which allow back-flow. We shall explain the components of the model and how the model works by applying it to a problem as an example.

The Lockers Problem

The new school has exactly 343 lockers numbered 1 to 343, and exactly 343 students. On the first day of school, the students meet outside the building and agree on the following plan. The first student will enter the school and open all the lockers. The second student will then enter the school and close every locker with an even number. The third student will then 'reverse' every third locker; i.e. if the locker is closed, he will open it, and if the locker is open, he will close it. The fourth student will reverse every fourth locker, and so on until all 343 students in turn have entered the building and reversed the relevant lockers. Which lockers will finally remain open?

Understand the Problem

First, one must realize that such a scenario could hardly happen in real life, simply because it is not possible for 343 kids to wait patiently for their turn to do such a tedious job! Thus, the problem is one of 'fantasy' or 'cognitive playfulness'. The key parts of the problem to take time to understand would be:

- that to 'reverse' a locker means to open it when it is closed and to close it when it is open;
- which lockers would a particular student act on.

It is common to see students staring at a blank piece of paper when they are stuck at a problem. The problem is too difficult for him or sometimes even for him to understand and so he cannot start. It is important that something is done and the following describes how *heuristics* (meaning: serving to find out or discover, *Oxford English Dictionary*) can help start the problem-solving process.

Two useful *heuristics* to help understand the problem are to *act it out* and to *consider a simpler problem (smaller numbers)*. Consider the problem when there are only 10 students. We list quickly the 10 lockers as 1, 2, ..., 10 on a piece of paper. Now we act out what the 10 students will do, using the following *representation*: a backslash '/' for open and a slash '\' over the backslash to obtain a cross '×' for close. The following shows what will result after a few students have been acted out (the actions taken in each step are in bold):

1)   1   2   3   4   5   6   7   8   9   10
      /   /   /   /   /   /   /   /   /   /

2)   1   2   3   4   5   6   7   8   9   10
      /   ×   /   ×   /   ×   /   ×   /   ×

3)   1   2   3   4   5   6   7   8   9   10
      /   ×   ×   ×   /   ×/ /   ×   ×   ×

4)   1   2   3   4   5   6   7   8   9   10
      /   ×   ×   ×/ /   ×/ /   ×/ ×   ×

The problem (not the solution) looks clear now. If we want to, we could solve the simplified problem for 10 students. This may be our first plan.

Note: Schoenfeld expands Pólya's model in his book *Mathematical Problem Solving* and suggests the following to try to understand the problem.

> Transcribe the problem in shorthand, drawing a diagram or making a rough sketch if you can. If someone took away the original problem statement, would you have enough to work with? Have you really identified all the essential information in the problem statement?

Devise a Plan

A plan is vital to the success of any endeavour. It has been said, "He who fails to plan, plans to fail."

It is common, when asking a student who is stuck at a problem what he is doing, to receive the reply "I don't know." Even though he may be scribbling something on the paper (which is usually better than nothing), there is no plan driving his effort. It is important for the problem solver to be able to articulate his plan so that he is guided by it.

Back to our problem, our first plan is: we will *consider a simpler problem (smaller numbers)* for 10 students, and *look for patterns.*

Carry out the Plan
With the plan decided, carrying it out requires the skills and knowledge that should be automatic in the problem solver. If there is a severe lack of resources, such as content knowledge, algebraic skills or mathematical techniques, the best plan will go awry. Thus, it is very important in teaching mathematical problem solving to know that a sufficiently strong foundation in mathematics knowledge is often necessary for success. In this case, the student must know the concept of multiples of numbers very well for efficiency.

So let us continue:

5)   1   2   3   4   5   6   7   8   9   10
      /   ×   ×   ×/ ×   ×/ /   ×/ ×   ×/

6)   1   2   3   4   5   6   7   8   9   10
      /   ×   ×   ×/ ×   ×× /   ×/ ×   ×/

7)   1   2   3   4   5   6   7   8   9   10
      /   ×   ×   ×/ ×   ×× ×   ×/ ×   ×/

8)   1   2   3   4   5   6   7   8   9   10
      /   ×   ×   ×/ ×   ×× ×   ×× ×   ×/

9)   1   2   3   4   5   6   7   8   9   10
      /   ×   ×   ×/ ×   ×× ×   ×× ×/ ×/

10)

	1	2	3	4	5	6	7	8	9	10
	/	×	×	×/	×	××	×	××	×/	××

Thus we have a solution for our simplified problem as 'Lockers 1, 4 and 9 will remain open at the end'.

It will take different lengths of time for different persons to be able to find a pattern but the pattern once found has to be stated as a conjecture. It is very important to tell students that in mathematics, pattern is NOT proof.

Our conjecture after carrying out the initial plan is:
The lockers that will remain open will be those whose number is a square, i.e. when there are 343 students, they are 1, 4, 9, 16, 25, 36, 49, ..., 324.

Check
Returning to the original problem, we check our conjecture for the next square that we did not actually act out, i.e., 16. We note that we need only to look at the first 16 lockers (out of the original 343) since students numbered 17 and above will not touch the first 16 lockers, in general Student n will not touch lockers numbered less than n. We do so with the following result:

1	2	3	4	5	6	7	8	9	10	11	12	13
/	×	×	×/	×	××	×	××	×/	××	×	×××	×

14	15	16
××	××	××/

The conjecture that the squares will remain open looks good so far!

Understand the Problem 2
We go back to Stage I to try to understand the problem deeper. Can we make progress by asking the question:

- What feature(s) of a square causes the corresponding locker to end up open? (Note that this is still a conjecture.)

The answer may not be obvious and so we ask a simpler question:

- By *acting it out* slowly, why does a particular locker end up open (or closed)?

By *acting it out* or by looking closely at the *representation* above, one will eventually realize that if a locker is 'touched' an odd number of times, it will end up open; if it is 'touched' an even number of times, it will end up closed.

Next, we use the heuristic *restate the problem in another way*.

- What type of number causes the corresponding locker to be touched an odd number of times?

Returning to the idea of features of a square, we ask:

- What feature(s) of a square causes the corresponding locker to be touched an odd number of times? (Note that this is still a conjecture.)

The series of questions involving the conjecture of squares, the lockers being touched an odd number of times, features of a square, distills to a more fundamental question which puts away the conjecture of squares for the moment.

- What feature(s) of a number causes the corresponding locker to be touched?

Again this may need different lengths of time for different persons. Eventually, we reach the understanding that a locker is touched only by students whose numbers are factors of the number of the locker. Thus, the number of times a locker is touched is exactly the number of the factors of its corresponding number.

Restating the problem in another way again:

- What type of number has an odd number of factors?

The series of questions using heuristics which twists, turns and focuses the original problem is crucial to better understand the problem before formulating a new plan of attack.

Remark: In some sense, the series of questions arise naturally and reflect the sequence of thinking that, hopefully, narrows down the original problem to one that can be solved.

Devise a Plan 2
Plan 2 will now be to count the number of factors of some numbers and try to understand why a square has an odd number of factors while the others have even numbers. We will *look for a pattern*.

Carry out the Plan 2
We will count the number of factors for 1 to 10 and then a few other random numbers when we spot a pattern. *Listing the factors systematically* will be helpful.

Number	Number of factors	Number	Number of factors
$1 = 1$	1	$7 = 1 \times 7$	2
$2 = 1 \times 2$	2	$8 = 1 \times 8 = 2 \times 4$	4
$3 = 1 \times 3$	2	$9 = 1 \times 9 = 3 \times 3$	3
$4 = 1 \times 4 = 2 \times 2$	3	$10 = 1 \times 10 = 2 \times 5$	4
$5 = 1 \times 5$	2	$16 = 1 \times 16 = 2 \times 8 = 4 \times 4$	5
$6 = 1 \times 6 = 2 \times 3$	4	$343 = 1 \times 343 = 7 \times 49$	4

It is now quite clear that a square has an odd number of factors while a non-square has an even number of factors – every factor of a number has a distinct 'partner', except for the root of the perfect square whose 'partner' is itself. At this point, we have essentially solved the problem as we now have the means—look for squares that have an odd number of factors—to find out if a locker will be open.

Devise a Plan 3

We will now write out a 'clean' solution focusing on showing that only the squares have an odd number of factors. We will do this by showing that the factors come in pairs.

Carry out the Plan 3

Let n be a positive integer. Let a_1, a_2, ..., a_k be all the k distinct factors of n with $a_1 < a_2 < ... < a_k$. Observe the following:

(i) n is the product of the i-th smallest factor a_i and the i-th largest factor a_{k+1-i}, i.e. $n = a_i a_{k+1-i}$, where $a_i \le a_{k+1-i}$.

(ii) If n is not a square, for $n = a_i a_{k+1-i}$, we have $a_i < \sqrt{n}$ and $a_{k+1-i} > \sqrt{n}$. This implies that each factor has a 'partner' factor distinct from itself such that the product of the two factors is n. Hence the number of factors must be even.

(iii) If n is a square, then $\sqrt{n}$ is an integer. As in (ii), each factor less than $\sqrt{n}$ has a 'partner' factor distinct from itself such that the product of the two factors is n. Observe that $n = \sqrt{n} \times \sqrt{n}$ and so this factor $\sqrt{n}$ has no distinct 'partner'. Hence the number of factors must be odd.

Now each locker will be 'touched' by a student whose number is a factor of the number of the locker. A square will be touched by an odd number of students because it has an odd number of factors as shown above. In the sequence of 'open' followed by 'closed', an odd number of actions will end in the locker eventually open. A non-square will, by the result above, be closed at the end.

Hence, the lockers that will remain open will be those whose number is a square, i.e. when there are 343 students, they are 1, 4, 9, 16, 25, 36, 49, ..., 324.

Check and Expand

We can check the solution for the locker 25. *Act out* the visitors to Locker 25, and as before make a backslash '/' for open and a slash '\'

over the backslash to obtain a cross '×' for close: 25 has factors 1, 5, 25 and so we have ×/, which means 'open'.

We are done for this problem but not done with the problem-solving process. It is a key feature of the model that the solver should try to 'expand' the problem even though it is solved. By expanding, we mean one of the following:

- finding other solutions which are 'better' in the sense of elegance, succinctness, or with a wider applicability
- posing new problems
  - *adapting* by changing certain features of it (eg. change some numbers, change some conditions, consider the converse)
  - *extending* to problems which are more 'difficult' or which have greater scope
  - *generalizing* to problems which would include the given problem as a special example

We sketch an alternative solution that uses prime factorisation to prove that a natural number is a square if and only if it has an odd number of factors as follows.

Clearly, 1 has an odd number, 1, of factors.

Let $n > 1$ be a natural number and $n = p_1^{k_1} p_2^{k_2} \cdots p_s^{k_s}$ be its prime factorization. Let m be a factor of n. Then $m = p_1^{l_1} p_2^{l_2} \cdots p_s^{l_s}$ where $0 \leq l_i \leq k_i$ for all $i = 1, 2, \ldots, s$.

By the Multiplication Principle, the number of possible m's, i.e., factors of n is $(k_1+1)(k_2+1)\ldots(k_s+1)$. This product is odd if and only if each of its factors (k_i+1) is odd, which in turn happens if and only if each k_i is even. Thus the number of factors of n is odd if and only if n is a square.

We pose the following problems with sketches of their corresponding proofs.

<u>Adaptation 1</u>:   The i-th student reverses every i-th locker except Locker i.

<u>Sketch of solution</u>:   Now, every locker is touched as many times as its number of factors less one. Thus, a non-square is touched an odd number of times and so remains open at the end.

<u>Adaptation 2</u>:   The i-th student reverses every locker whose number is a factor of i.

<u>Sketch of solution</u>:   The multiple-factor relationship in the original problem is now reversed. Thus, we are interested in the number of multiples of a number m. The locker with number m will remain open at the end if and only if $\left\lfloor \frac{343}{m} \right\rfloor$ is odd.

<u>Extension</u>:   There are m stages to finally open the locker (for example: put the key in the lock, turn the key, pull down the handle, etc.). The i-th student goes to every i-th locker and acts out the next stage, and if the locker is fully open, closes and locks it. Which lockers remain open in the end?

<u>Sketch of solution</u>:   Again, every locker is touched as many times as its number of factors but now, it remains open if and only if its number of factors is m mod $(m+1)$.

<u>Generalisation 1</u>:   There are n lockers. How many lockers remain open?

<u>Sketch of solution</u>:   The lockers that remain open are the squares which do not exceed n. Number of lockers that remain open = $\left\lfloor \sqrt{n} \right\rfloor$.

1.3 Focusing with Schoenfeld's Framework

We continue with Schoenfeld's comments in his preface to *Mathematical Problem Solving*.

> The next day I spoke to the colleague who trained our
> department's team for the Putnam exam, a prestigious

nationwide mathematics competition. Did he use Pólya's book? "No," he said. "It's worthless." His teams did quite well, so there must have been some truth in what he said.

Schoenfeld grappled with the apparent worth of Pólya's model and the real-world failure of its application in the classroom. His research led him to realize that there was more than just a direct application of the model and that in fact, other factors are crucial in successful problem solving. His research culminates in the construction of a framework for the analysis of complex problem-solving behaviour. The framework, in his words, consists of four aspects:

1.	Cognitive resources – the body of facts and procedures at one's disposal
2.	Heuristics – 'rules of thumb' for making progress in difficult situations
3.	Control – having to do with the efficiency with which individuals utilise the knowledge at their disposal
4.	Belief systems – one's perspectives regarding the nature of a discipline and how one goes about working on it

With regard to our attempts to teach problem solving, Schoenfeld's framework makes us realize that Pólya's model or the teaching of heuristics is not all there is. The problem solver needs also to address the availability of resources and exercise cognitive and affective control over the problem solving process. The framework allows teachers to have a structure through which they can focus when difficulties arise in solving a problem. We use our example to illustrate how Schoenfeld's framework can be used.

Cognitive resources
With respect to the original locker problem, failure will result if there is a lack of knowledge of any of these: 'reverse' or 'every fourth' in *Understanding the Problem*; 'multiples' and 'factors' in *Devise a Plan* and *Carry out the Plan*; 'prime factorisation' for the alternative proof in *Check and Expand*.

Our study of problem solving (Quek, Choy, Dong, Toh & Ho, 2007) suggests that we regard *resources* as necessary but not sufficient for success in problem solving. Further, we recommend that teachers should not neglect the provision of *resources* during the planning of curriculum and the teaching of mathematics. It is not viable—at least in the school situation they were in—to devote a substantial amount of curriculum time meant for important content and concepts of mathematics and their interrelationships to just teach processes, i.e. Control and Heuristics, or 'motivate' students, i.e. change Beliefs.

Heuristics
Many problem solving programmes emphasise the use of heuristics. Schoenfeld is no exception. The use of heuristics is the most visible facet of successful problem solving. For the observer, it would seem that the application of a suitable heuristic unlocks the puzzle. Schoenfeld initially saw the centrality of heuristics in Pólya's model. Later, he realized that the other three less visible aspects of resources, control and beliefs must be taken together to obtain a fuller picture of successful and unsuccessful problem solving.

In our example, heuristics are used to explore the problem (*act it out* and *consider a simpler problem (smaller numbers)*), as the basis of a plan (*consider a simpler problem (smaller numbers)* and *look for patterns*), and while carrying out the plan (*use a suitable representation (table)*).

A list of heuristics that we will use in this book is:
- restate the problem in another way
- think of a related problem
- work backwards
- aim for sub-goals
- divide into cases
- use suitable numbers (instead of algebra)
- consider a simpler problem
  - smaller numbers
  - special case - tighten conditions
  - fewer variables

- consider a more general case - loosen conditions
- act it out
- guess-and-check
- make a systematic list
- make a table
- look for patterns
- use equations/algebra
- draw a diagram
- use a suitable representation
- use suitable notation

Control

Control, also called metacognition, is a major determinant of problem-solving success or failure (Schoenfeld, 1985). In our opinion, much of the difficulty in teaching problem solving or any other aspect of mathematics in our schools arises from underemphasizing the role of metacognition. The teaching of heuristics *per se* often takes place in place of teaching problem solving with all its different aspects. To be fair, the teaching of control during problem solving is very difficult.

In our example, control is needed in the second round of understanding the problem where the series of questions asked finally focused on the key aspect of factors. The good problem solver will 'sense' that he cannot proceed with a solution until he is sure of why the lockers with square numbers remain open at the end. He will 'force' himself to apply some heuristics to make sure that the problem is well understood before rushing on to work on it.

Control is needed when devising a plan. Again, the good problem solver is patient and willing to go through more than one plan, choosing a modest goal first before launching into a full scale attack on the problem. Many students with poor control end up spending more time looking for a quick solution – to them, 'quick' means a 'magic bullet' of a formula that will give the answer.

Control is needed when carrying out the plan, to decide on a suitable representation (the slash and cross) in marking out the actions of opening and closing in Plan 1, and to decide how many operations to continue doing in a search for a pattern in Plan 2. Control is constantly needed to monitor if we are still on the right track when algebraic manipulations and numerical calculations are invólved. In fact, control is necessary to regulate the starts and stops, turns and returns of the problem solving journey.

Even good students do not usually want to extend a problem, being satisfied with the successful solving of the original problem. Such students fall just short of developing a mature mathematical way of thinking. Control is needed when checking and extending. In some sense, this control is for the larger arena of *all* problems in that the insights gained from following through with Pólya's fourth and last stage will 'save time' for future problems. In our example, the adaptations offer reinforcement for the method used and the alternative solution offered a number-theoretical perspective.

Control also means knowing when to store a problem for another time, or to give it up permanently or temporary, to regurgitate at a suitable time. For some students, we may encourage them to take even days to work out the relevant questions of Understanding the Problem 2.

Beliefs

If a person believes that real life does not need mathematics or at best, that mathematics should be left to a few mathematicians, then it will not be surprising that, that very person will not find it of any value to solve 'difficult' mathematics problems. If a person believes that there is always a fixed method to solve mathematics problems and anything else beyond that belongs to the realm of the nerds (again), then such a person will give up whenever he encounters a mathematics problem that cannot be solved within a short time.

Beliefs are strong and ingrained in a person and beliefs about mathematics strongly affects how one approaches mathematics and mathematics problems.

In our example, a person who believes that mathematics problems are to be solved 'in a short period of time or not at all' will give up quickly when acting out the opening and closing of the lockers. Again, a person who believes that Mathematics is all about formula may likely give up after realizing that there is no 'magic bullet' formula for this problem.

A person who knows that listing is a worthwhile heuristic to explore a problem or to solve part of it perseveres even though the listing may seem slow, tedious and to reward little. A different person may believe that 'listing is like a train - it starts slowly but picks up speed' and thus, perseveres knowing that the initial cases though slow do not translate into proportional time when extended to the general case. Thus, after spending quite some time slowly recording the first few students acting on the lockers, the rest would be a breeze.

In summary, successful problem solving is not a mere application of some model. It is the complex interaction of at least four components (Resources, Heuristics, Control and Beliefs). On the flip side, lack of success in problem solving can be attributed to a haphazard approach lacking one or more of the positive aspects of the said components.

1.4 The Practical Paradigm

1.4.1 Evolution of Problem Solving instruction

As we taught some problem solving classes, we noticed that students were resistant to following the stages of Pólya's model. Although we coaxed and encouraged, generally, students did not take to the model. Even the high ability mathematics students who could solve the given problems refused to make the extra effort to finally Check and Expand the problem.

For them, although there was apparent progress in their ability to use heuristics, the metacognitive part of the problem solving process still left much to be desired. Schoenfeld (1985) asked out loud specific control questions at regular intervals as a way to make students pause and be more aware of their thinking.

Sometimes, it is good to steal a sideways glance at what others are doing. Tay (2001) discussed the use of Cloze and comprehension passages to teach and assess the skills of reading and writing mathematics. The paradigm for this came from the teaching of language. In an attempt to 'make' the students follow the Pólya model, especially when they were clearly struggling with the problem, we decided to construct a worksheet (described below) like that used in science practical lessons and told the students to treat the problem solving class as a mathematics 'practical' lesson. In this way, we hoped to achieve a paradigm shift in the way students looked at these 'difficult, unrelated' problems which had to be done in this 'special' class. The science practical lesson is very much accepted by students as part of science education and many have an understanding that it is to teach them how to 'do' science. Woolnough and Allsop (1985) stated clearly what is to be achieved in science education.

> As we look at the nature of science we see two quite distinct strands. The knowledge, the important content and concepts of science and their interrelationships, and also the processes which a scientist uses in his working life. In teaching science we should be concerned both with introducing students to the important body of scientific knowledge, that they might understand and enjoy it, and also with familiarizing students with the way a problem-solving scientist works, that they too might develop such habits and use them in their own lives.
>
> (Woolnough and Allsop, 1985, p.32)

It is instructive to see that we could just replace 'science' with 'mathematics' and the preceding passage reads just as truly. Practical work to achieve the learning of the scientific processes has a long history of at least a hundred years and can be traced to Henry Edward Armstrong (Woolnough and Allsop, 1985; Armstrong, 1891). Despite much debate of how exactly it is to be carried out (Woolnough and Allsop, 1985), practical work is accepted as a mainstay in science education. It is certainly conceivable that similar specialised lessons and materials for mathematics may be necessary to teach the mathematical processes, including and via problem solving. To this, we propose the idea of Mathematics Practical.

Thus, in a 55-minute Mathematics Practical lesson (we assume that each lesson has a duration of 55 minutes), the class focuses on one problem and students have to work this out on a Mathematics Practical worksheet.

The worksheet consists of 4 pages (see Appendix 1), each page corresponding to one Pólya stage. Ideally, the student will follow the model and go through all 4 stages, with suitable loopbacks (work on blank piece of paper and attach at the relevant place).

However, so as not to straitjacket an unwilling problem solver, the student may jump straight to Stage 3: Carry out the Plan and is given up to 15 minutes to solve the problem there. A student who completes the problem in time needs only do the last Pólya stage. If not, it is explained to him that the Pólya model would allow him to do better than what he has achieved and so he is required to go through all four Pólya stages.

These alternatives are intended to differentiate between the needs of the different problem solvers, where the unsuccessful student is required to systematically and metacognitively go through the Pólya model and the successful student is allowed to leapfrog to the fourth stage which is important for consolidation of the method and gaining a fuller insight to the problem.

Certainly, a particular student may have to go through all 4 stages for one problem and not for another problem where the plan to him is obvious and he needs only to Check and Expand. The choice allowed here is to show that explicit use of Pólya's model is not necessary all the time but would be very useful when one is stuck and when one is learning to problem-solve.

Appendix 1 is a template for the Practical Worksheet and Appendix 2 is a scan of work done on the Lockers Problem using the Practical Worksheet by a graduate of the problem solving module.

1.4.2 Choosing problems for the Mathematics Practical

The successful implementation of any problem solving curriculum hinges on the choice of *appropriate* problems. However, *appropriate* problems may have different connotations for different people.

The nurturing of problem solving skills requires students to solve meaningful problems. Lester (1983) claimed that posing the cleverest problems is not productive if students are not interested or willing to attempt to solve them. The implication is that mathematical problems have to be chosen judiciously. What should be the criteria for choosing good problems? Problems selected for a course must satisfy five main criteria (Schoenfeld, 1994, cited in Arcavi, Kessel, Meira, & Smith, 1998, pp. 11-12):

1. Without being trivial, problems should be accessible to a wide range of students on the basis of their prior knowledge, and should not require a lot of machinery and/or vocabulary.
2. Problems must be solvable, or at least approachable, in more than one way. Alternative solution paths can illustrate the richness of the mathematics, and may reveal connections among different areas of mathematics.
3. Problems should illustrate important mathematical ideas, either in terms of the content or the solution strategies.
4. Problems should be constructible or solvable without tricks.
5. Problems should serve as first steps towards mathematical explorations, they should be extendable and generalizable; namely, when solved, they can serve as springboards for further explorations and problem posing.

Conscious of the fact that the choice of problems was critical in our problem solving project, we were guided by the following principles: (1) the problems were interesting enough for most if not all of the students to attempt the problems; (2) the students had enough "resources" to solve the problem; (3) the content domain was important but subordinate to processes involved in solving it; and (4) the problems were *extendable* and *generalizable*.

Guided by these criteria, we have chosen problems for the Mathematics Practical lessons (Chapter 4). Some of these choices of the problems were based on data collected from previous students solving these problems. The authors are in the process of developing three additional sets of corresponding mathematics problems in Chapter 4. These problems will be available in the near future.

1.4.3   Scaffolding in the Mathematics Practical

What does the teacher do when students are working on a problem using the Practical Worksheet? As in the science practical where the teacher or technician provides help in setting up apparatus and in understanding the experimental setup and analysis with the view to learn the processes of science, the mathematics teacher should be clear that his overt help is towards improving the problem solving processes of the student seeking help.

To this end, we propose three levels of help that a teacher can give to a student. The levels are hierarchical and one level should be given only after an earlier level has failed. In the crucial level which we call Level 0, we emphasise the student learning and reinforcing the Pólya model. We may ask the student if he knows what Pólya stage he is in, and what would one normally do in such a stage. We help by asking the two Control questions. In Level 1, we suggest *specific* heuristics to get the work moving. Level 2 is to be avoided as much as possible and is included only for the important aspect of ensuring that the self-esteem of the student is not seriously damaged by his perceived failure and helplessness on the problem. Here, we give problem specific hints, which essentially is a throwback to the 'usual' help afforded by mathematics teachers. The objective of the practical paradigm is for students to internalize Level 0, and ask for Level 1, and to a much lesser extent Level 2, hints only when pressed for time.

Level	Feature	Examples based on the Lockers Problem
0	Emphasis on Pólya stages and control	What Pólya stage are you in now? Do you understand the problem? What exactly are you doing? Why are you doing that?
1	Specific heuristics	Why don't you try with fewer lockers (*use smaller numbers*)? Try *looking for a pattern*.
2	Problem specific hints	Think in terms of the locker rather than the student – what numbers get to touch the locker?

The set of problems in Chapter 4 are placed within the template of the Practical Worksheet and include suggestions for scaffolding the students. Do take note that Level 0 prompts are not specifically stated. The teacher must always remember to first ask Level 0 questions; he may also direct the student to respond to the instructions and questions in the Practical Worksheet template as they are essentially Level 0 questions.

1.4.4 Assessment of the Practical Worksheet

Meeting the challenge of teaching mathematical problem solving to students, calls for a curriculum that emphasises the process (while not neglecting the product) of problem solving and an assessment strategy to match it, to drive teaching and learning. Effective assessment practice "begins with and enacts a vision of the kinds of learning we most value for students and strive to help them achieve" (Walvoord & Anderson, 1998).

Correspondingly it is common knowledge that most students will not study for curricular components which are not assessed. We have seen how the Practical Worksheet fits into the paradigm of a Mathematics Practical. We now present the scoring rubric as an assessment instrument for the Practical Worksheet.

The scoring rubric focuses on the problem-solving processes highlighted in the Practical Worksheet. There are four main components to the rubric, each of which would draw the students' (and teachers') attention to the crucial aspects of as authentic as possible an attempt to solve a mathematical problem:

- Applying Polya's 4-phase approach to solving mathematics problems
- Making use of heuristics
- Exhibiting 'control' during problem solving
- Checking and expanding the problem solved

In establishing the criteria for each of these facets of problem solving, we ask the question, "What must students do or show to suggest that they have used Polya's approach to solve the given mathematics problems, that they have made use of heuristics, that they have exhibited 'control' over the problem-solving process, and that they have checked the solution and extended the problem solved (learnt from it)?"

The rubric is presented below. The total score that a student is able to score for a problem is 20 marks.

- *Polya's Stages* [0-7] – this criterion looks for evidence of the use of cycles of Polya's stages (Understand Problem, etc).
- *Heuristics* [0-7] – this criterion looks for evidence of the application of heuristics to understand the problem, and devise/carry out plans.
- *Checking and Expanding* [0-6] – this criterion is further divided into three sub-criteria:
    - Evidence of checking the correctness of solutions [1]
    - Providing for alternative solutions [2]
    - Adapting, extending and generalizing the problem [3] – full marks for this are awarded for one who is able to

> provide (a) two or more generalizations of the given
> problem with solutions or suggestions to solution, or (b)
> one significant extension with comments on its
> solvability.

The rubric is designed to encourage students to go through the Polya
stages when they are faced with a problem, and to use heuristics to
explore the problem and devise a plan. They would return to one of the
first three phases (see the Practical Worksheet in Appendix 1) upon
failure to realize a plan of solution. Students who 'show' control
(Schoenfeld's framework) over the problem solving process earn marks.
For example, a student who did not manage to obtain a correct solution
would be able to score up to five marks each for *Polya's Stages* and for
Heuristics, making a total of ten, if they are show evidence of cycling
through the stages, use of heuristics and exercise of control.

The rubric allows the students to earn as many as 70% of the total 20
marks for a correct solution. However, this falls short of obtaining a
distinction (75%) for the problem. The rest would come from the marks
in *Checking and Extending*. Our intention is to push students to Check
and Expand the problem, an area of instruction in problem solving that
has been largely unsuccessful (Silver, Ghousseini, Gosen, Charalambous,
& Strawhun, 2005).

Appendix 3 is a template of the Assessment Rubric.

References

Arcavi, A., Kessel, C., Meira, L., & Smith, J. (1998). Teaching
     mathematical problem solving: A microanalysis of an emergent
     classroom community. In A. Schoenfeld, E. Dubinsky, & J. Kaput
     (Eds.), Research in Collegiate Mathematics Education III (pp. 1-70).
     Providence, RI: American Mathematical Society.

Armstrong, H. E. (1891). *The teaching of scientific method.* London:
     Macmillan.

Lester, F. K. (1983). Trends and issues in mathematical problem-solving research. In R. Lesh & M. Landau (Eds.), *Acquisition of mathematics concepts and processes* (p. 229-261). Orlando, FL: Academic Press.

Pólya, G. (1954). *How to solve it*. Princeton: Princeton University Press.

Quek, K. S., Tay, E. G., Choy, B. H., Dong F. M., Toh T. L., & Ho F. H., Mathematical Problem Solving for Integrated Programme Students: Beliefs and performance in non-routine problems. *Proceedings EARCOME 4 2007: Meeting the Challenges of Developing a Quality Mathematics Education Culture* (2007) 492-497.

Schoenfeld, A. (1985). *Mathematical problem solving*. Orlando, FL: Academic Press.

Silver, E. A., Ghousseini, H., Gosen, D., Charalambous, C., & Strawhun, B. T. F. (2025). Moving from rhetoric to praxis: Issues faced by teachers in having students consider multiple solutions for problems in the mathematics classroom. *Journal for Mathematical Behavior*, 24, 287 – 301.

Tay, E. G. (2001). Reading Mathematics. *The Mathematics Educator* 6(1) 76-85.

Walvoord, B.E., Anderson, V.J. (1998). *Effective grading: a tool for learning and assessment*. California: Jossey-Bass Publishers.

Woolnough, B. & Allsop, T. (1985). *Practical work in science*. (Cambridge Science Education Series) Cambridge: Cambridge University Press.

Chapter 2

Scheme of Work and Assessment of the Mathematics Practical

2.1 Scheme of Work

The first part of this chapter provides the readers with a proposed set of Scheme of Work for the Mathematics Practical Lessons. The Scheme of Work provides the readers with the entire structure of the Mathematics Practical Lessons. The detailed lesson plan is presented in Chapter 3.

The problem number referred to in the fourth column in the table of the Scheme of Work refers to the problem number in Chapter 4.

Throughout all the Mathematics Practical Lessons,

- the students attending the problem solving course should be organized to work in pairs for all the problems that are done in class;
- the teacher should attempt to keep to the time allocation in lesson plan in Chapter 3; and
- during seatwork, teacher should elicit and attends to student questions.

Lesson	Theme/ Topic	Specific Instructional Objectives	Suggested Task & Activities	Problem No.
1	What is a problem?	Student should be able to: • differentiate between an exercise and a problem in mathematics • identify the strategies and dispositions favourable to successful problem solving	• Student will work on 2 questions to enable him to experience the mathematical and psychological dimensions of problem solving of the non-routine kind • Teacher will distinguish a problem from an exercise by defining what a problem is • Teacher will model the solution of a problem	1

Lesson	Theme/ Topic	Specific Instructional Objectives	Suggested Task & Activities	Problem No.
2	Pólya's problem solving strategy	Student should be able to: • identify the phases typically found in solving non-routine mathematics problems • state the key features in each phase of problem of solving • describe the actions of a problem solver during each phase	• Teacher will model the strategy by working on the homework problem • Student will work on a problem using the model – introduction of Problem of the Day (POD) • First page of Practical Worksheet to be used for POD • Teacher will solve POD using the model • Heuristics – *Consider a simpler problem (smaller numbers), Make a systematic list, Look for patterns*	2, 3

Lesson	Theme/ Topic	Specific Instructional Objectives	Suggested Task & Activities	Problem No.
3	Using heuristics to understand the problem	Student should be able to: • state what are heuristics used for • list a set of heuristics • use heuristics to get started on solving a problem	• Teacher will explicitly mention some heuristics by name when reviewing the homework • Teacher will explain the meaning of the word *heuristics* and list some examples • Teacher will put up a list of heuristics on the notice board • POD • First 3 pages of Practical Worksheet to be used for POD • Teacher will highlight the usefulness of heuristics to understand the POD • Heuristics – *Act it out, Draw a diagram, Restate the problem in another way, Aim for sub-goals*	4, 5

Lesson	Theme/ Topic	Specific Instructional Objectives	Suggested Task & Activities	Problem No.
4	Using heuristics for a plan	Student should be able to: • identify the role of resources in problem solving • use heuristics for a plan	• Teacher will highlight the use of heuristics in reviewing the homework • Teacher will explain the use of heuristics in Stages 2 and 3, and explain that heuristics do not replace resources but the two are complementary • POD • First 3 pages of Practical Worksheet to be used for POD • Heuristics – *Act it out, Restate the problem in another way, Consider a simpler problem (smaller numbers), Use suitable numbers (instead of algebra), Think of a related problem, Divide into cases*	6, 7

Lesson	Theme/ Topic	Specific Instructional Objectives	Suggested Task & Activities	Problem No.
5	The Mathematics Practical	Student should be able to: • appreciate the rationale of the worksheet • use the practical worksheet to record their attempts at solving problems	• Teacher will highlight the use of heuristics in reviewing the homework • Teacher will explain the practical paradigm • Teacher will explain the purpose of the practical worksheet and demonstrate how to fill it in while working on a problem • Heuristics – *Act it out, Consider a simpler problem (smaller numbers), Look for patterns, Use a suitable representation, Restate the problem in another way*	8, 9

Lesson	Theme/ Topic	Specific Instructional Objectives	Suggested Task & Activities	Problem No.
6	Check and Expand	Student should be able to state the features of Check and Expand: • check the accuracy, the suitability and elegance of the solution • consider alternative solutions • adapt, extend or generalize the problem	• Teacher will highlight the use of the Practical Worksheet in reviewing the homework • Teacher will explain the features of looking back and looking forward, i.e., Check and Expand • Teacher will elaborate on adapt, extend and generalize using a previous problem • POD (Practical Worksheet) • Heuristics – *Use suitable numbers (instead of algebra), Think of a related problem*	10

Lesson	Theme/ Topic	Specific Instructional Objectives	Suggested Task & Activities	Problem No.
7	More on adapting, extending and generalizing	• Student should be able to pose at least one new problem from the POD	• Teacher will discuss posed problems from the homework • Teacher will reiterate the features of looking back and looking forward, i.e., Check and Expand • POD • Heuristics –*Use suitable numbers (instead of algebra), Think of a related problem*	11, 12

Lesson	Theme/ Topic	Specific Instructional Objectives	Suggested Task & Activities	Problem No.
8	Schoenfeld's Framework	Student should be able to: • describe what 'control' is about in problem solving • act on control instructions from the teacher: ○ What are you doing? ○ Why are you doing it?	• Teacher explains Schoenfeld's framework • Teacher will explain the need for control • Teacher will model Control in reviewing the homework • Teacher will act as external control during POD • Heuristics – *Act it out, Use suitable numbers (instead of algebra), Consider a simpler problem (smaller numbers), Look for patterns, Guess-and-Check, Divide into cases, Work backwards*	13, 14

Lesson	Theme/ Topic	Specific Instructional Objectives	Suggested Task & Activities	Problem No.
9	More on Control	Student should be aware of: • reaching a 'stuck' state; and • deciding on whether to: ○ carry on with the plan; or ○ abandon the plan; or ○ put on hold and try another plan Student will become aware of the affective dimensions of problem solving	• Teacher will model Control in reviewing the homework • Teacher will explain what it is to be 'stuck', and making a decision to move out of the 'stuck' situation Teacher will highlight the danger of 'fixation' • POD – student will write down decision making statements in the prescribed column of Stage III in the practical worksheet • Heuristics – *Consider a simpler problem (special case – tighten conditions,) Draw a diagram, Think of a related problem*	15, 16

Lesson	Theme/ Topic	Specific Instructional Objectives	Suggested Task & Activities	Problem No.
10	Revision	Student should be able to answer the following: • What is a problem? • Draw the model of Pólya's problem solving strategy. • What are the heuristics? Name some of them. • Name the 4 components of Schoenfeld's framework. What do you do when you are 'stuck'?	• Teacher will review homework • Teacher will revise key features of problem solving by discussing the module and asking students questions • POD • Heuristics - *Solve a simpler problem (smaller numbers), Look for patterns, Use equations/algebra, Think of a related problem, Divide into cases*	17
11	End of module assessment		• Practical worksheet – a 50-minute problem	

2.2 Assessment Structure

We propose an assessment for the Mathematics Practical as follows: 40% of the total assessment consists of a final practical test and 60% comprises continual assessment.

	Practical Test	Continual Assessment
Weighting	40%	60%
Details	1 problem to be done on a practical worksheet in 50 minutes.	2 pairwork POD (30%): Teacher selects one for the whole class; student selects his best 2 take-home individual homework (30%): Teacher selects one for the whole class; student selects his best (30%)

Chapter 3

Detailed Lesson Plans

3.1 Introduction

This chapter provides the detailed lesson plans of all the 10 Mathematics Practical lessons. The duration of each lesson is 55 minutes. The problems selected for each segment of the lesson and the suggested time for each part of the activities is indicated. Each problem selected for each lesson or part of the lesson is designed to meet the specific instructional objective detailed in the Scheme of Work in Chapter 2. Teachers are strongly advised to keep to the problems and the suggested time when they first conduct the Mathematics Practical lessons.

Lesson 1: What is a Problem?

Exercise and Problem: Jugs (10 min.)
Teacher begins the lesson by asking their students to have a look at the two questions and attempt to solve them in pairs.
1. A jug holds 5 litres of water when full. How many jugs do we need to hold 47 litres of water?
2. You are given two jugs, one holds 5 litres of water when full and the other holds 3 litres of water when full. There are no markings on either jug and the cross-section of each jug is not uniform. Show how to measure out exactly 4 litres of water from a fountain.

Discussion: What is a problem? (8 min.)
Teacher explains the difference between an exercise and a problem as follows:
- No easy access to a procedure for solving the problem
- There must be a desire/motivation/reason to solve the problem
- Problem solving begins when an effort is made to find the solution

Teacher relates the 3 criteria to students' efforts on Questions 1 and 2.

Problem Solving: Teacher models (35 min.)

(3 min.)Teacher asks for student answers to Question 1. Teacher classifies the question as an exercise.

(5 min.)Teacher asks for student answers to Question 2. Ask a student with correct answer to explain her solution. Ask a student with incorrect answer to explain her solution too. Question them about their steps and make reference to them when possible during later explanation.

(7 min.) Teacher models the solution of the Question 2 by going through the first 3 stages of Pólya's model **without** explicitly mentioning Pólya. Teacher emphasises the need to understand the problem (no markings, exact), devising a plan (use some 'strategy' such as 'trial and error'), and carrying out the plan (while being aware if one should continue to pursue the plan).

(9 min.) Teacher checks the solution. Now teacher extends the problem: "What if the numbers change?" Ask students to work on the following in pairs:

   i.   Get 2 litres from 3 litre and 7 litre jugs.
  ii.   Get 6 litres from 12 litre and 16 litre jugs.
 iii.   Get 12 litres from 18 litre and 24 litre jugs.

(9 min.) Teacher asks students for solutions. Discuss why (ii) cannot be solved. Explain the idea of looking back and thinking of alternative solutions. Introduce Diophantine equation (an equation which admits only integer-valued solutions) for (i) as $3x + 7y = 2$. Show that an integer solution for the Diophantine equation is equivalent to the 'pouring in (positive term) and pouring out (negative term)'. Use the Diophantine equation to show that (ii) cannot be solved. Get students to solve (iii) using a Diophantine equation. (Show that there is more than one solution.)

Closure (2 min.)
Review what the criteria are for a question to be considered a problem. (Note that this and other things such as heuristics need to come up from time to time in all lessons just to refresh students' ideas.)

Impress upon the students that 'real' mathematical problem solving takes longer than 'normal' questions and involves strategies, resources and correct attitudes.

Tell students that this is the beginning of their journey into learning how to think like a mathematician.

Homework
Pose 3 questions based on the Jugs Problem and solve them.

Lesson 2: What is a Problem?

Review of Homework (10 min.)
Ask two volunteers to write down their 3 'jugs' questions on the board. Ask each of the two volunteers to solve the other's problems. Comment and discuss on the problems posed and the solutions. Ask the two students if they have gained a better understanding of the original problem by posing problems of their own.

Discussion: A model for problem solving (10 min.)
Teacher explains the advantages of having a model for problem solving:

- Know what to do when 'stuck'
- Systematic
- That is how mathematicians think

Teacher introduces George Pólya as a Hungarian mathematician who tried to help people solve mathematics problems. He wrote "How to solve it" and in it is his strategy of problem solving. Draw the diagram on the board, emphasizing the 'loopbacks'.

Pólya's Problem Solving Strategy

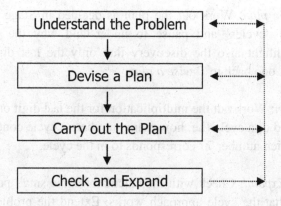

Problem of the Day (POD): Last Digit (20 min.)
Find the last digit of 13^{77}.

Teacher hands out to each student the first page of the Practical Worksheet for students to fill in Stage I: Understand the Problem.
Teacher walks around and asks these questions:
- Do you understand the problem?
- What is your plan to solve the problem?
- Did you check your work?
- Now that you have solved the problem, can you make up some new problems from this old one?

Discussion of POD (10 min.)
Ask students for the answer. Model problem solving on the board, emphasizing the Pólya stages as they are entered, left or returned to.

Understand the problem: Only the last digit is required. The calculator is not useful because it shows only about 10 digits.

Devise a plan: Decide to use the heuristics *Consider a simpler problem (smaller numbers)*, *Make a systematic list* and *Look for patterns*.

Carry out the plan: Work out the plan as decided in Stage 2. Pause at discovery of 'cycles' and pause to understand why the cycles will happen. Highlight also the discovery that only the last digit need be considered. Loop back to *Devise a plan*.

Devise a plan: Work out the multiplication for the last digit only and find out the period of a cycle (i.e. how many numbers a cycle contains). Then work out which number 77 corresponds to in the cycle.

Check and Expand: Check with the calculator some small powers of 13 to confirm that the cycle approach works. Extend the problem to other numbers and then to a different number of last digits.

Closure (5 min.)
Review Pólya's problem solving strategy. Ask students what the actions of a problem solver are during each phase. State the following actions to reinforce or supplement students' answers. (Prepare slide earlier.)

Understand the problem: Read through the problem carefully. Think about what the final answer would look like – an equation? a diagram? a single digit answer? a set?

Devise a plan: "He who fails to plan, plans to fail." Problem solver decides on a first (and subsequent) plan(s) of attack. These plans may involve heuristics to help to discover aspects of the problem.

Carry out the plan: Work out the plan as decided in Stage 2. Pause at discoveries and adjust for new plans if necessary.

Check and Expand: Check the solution. Look back and examin.e the solution and see if it can be improved or whether it can be used to solve other kinds of problems. Extend the problem.

Homework
Find the last digit of $1962^{2009} + 2009^{1962}$.

Lesson 3: Using Heuristics to Understand the Problem

Review of Homework (7 min.)

Teacher models and solves the problem on the board. Use the heuristic *Solve part of the problem* to work out 1962^{2009} and 2009^{1962} separately before putting them together to get the answer. Extend quickly to the multiplication and the subtraction of the two numbers. The teacher may comment that division is not so straightforward.

Discussion: Heuristics (6 min.)

Explain that *Heuristics* is a Greek word that means 'serving to discover'. Emphasise that a heuristic is not a magic method but that it is something problem solvers do to *help* them make some headway in attacking the problem. Quite often, the use of heuristics will result in the problem solver *discovering* something new or important about the problem.

Put up the following list of heuristics but state that the list is meaningless until one uses the heuristics. Say that in this and subsequent lessons, the teacher together with the class will write the name of the problem beside a heuristic when it is used.

- restate the problem in another way
- think of a related problem
- work backwards
- aim for sub-goals
- divide into cases
- use suitable numbers (instead of algebra)
- consider a simpler problem
  - o smaller numbers
  - o special case - tighten conditions
- consider a more general case – loosen conditions

- act it out
- guess-and-check
- make a systematic list
- make a table
- look for patterns
- use equations/algebra
- draw a diagram
- use a suitable representation
- use suitable notation

Problem of the Day: Phoney Russian Roulette (25 min.)

Two bullets are placed in two consecutive chambers of a 6-chamber revolver. The cylinder is then spun. Two persons play a safe version of Russian Roulette. The first points the gun at his hand phone and pulls the trigger. The shot is blank. Suppose you are the second person and it is now your turn to point the gun at your hand phone and pull the trigger. Should you pull the trigger or spin the cylinder another time before pulling the trigger?

Teacher hands out to each student the first 3 pages of the Practical Worksheet for students to fill in Stage I: Understand the Problem and Stage 2: Devise a Plan. They are then to write their solution in Section 3: Carry out the Plan.

Teacher walks around and suggests heuristics, especially to understand the problem.

Discussion of POD (15 min.)

Ask students for their answers. Ask a student with the correct answer and the correct reasoning to explain his problem solving process.

Finetune student's solution so as to highlight the use of the heuristics *Act it out* and *Draw a diagram* to understand 'two bullets in two consecutive chambers'. Highlight the use of the heuristic *Restate the problem in another way* (Which is less likely to destroy your own phone?) to discover that probability is involved.

Write the name of the problem Write the name of the problem beside *Act it out*, *Draw a diagram* and *Restate the problem in another way* on the list of heuristics on the notice board.

Closure (2 min.)

Ask students for the meaning of the word *Heuristics*. Ask which heuristics were used by them in solving the Problem of the Day.

Homework: 4-D coincidence

'*1 in 100 million chance: 6904 wins top two prizes in 4-D draw*' was the title of a news report on page H1 of The Straits Times Thursday 28 June 2007. Is this probability correct?

(The 4-D is a popular Singaporean gambling game which gives 23 prizes for 4-digit sequences obtained from 0000 to 9999. Repeats are allowed.)

Lesson 4: Using Heuristics for a Plan

Review of Homework (7 min.)

Teacher models and solves the problem on the board. Use the heuristics *Act it out* and *Restate the problem in another way* to lead the students to understand that the problem is not to find the probability of a **particular** number appearing twice but of **any** number appearing twice. Use the heuristic *Consider a simpler problem (smaller numbers)* to work out the probability of a double when two ordinary dice are tossed. The required probability would be the probability of a double when two 10,000-faced dice are tossed. (Write the name of the problem beside the relevant heuristics on the notice board.)

Discussion: Heuristics and Resources (5 min.)

Highlight how *Restating the problem* helped in understanding the problem and *Solving a simpler problem* helped in solving the actual problem. Emphasise that it will take time to know more heuristics, to know when to use them, and to know how to use them. Write the name of the problem beside *Consider a simpler problem (smaller numbers)* and *Restate the problem in another way* on the list of heuristics on the front notice board.

Explain also that heuristics by themselves are not enough. One must know sufficient mathematics, i.e. have enough resources, to complement good heuristics. For example, if one did not know how to calculate the probability of a combined event, then understanding the problem and having a 'good' plan may not be enough.

Problem of the Day: Same last digit (30 min.)

Show that the integer n always has the same last digit as its fifth power n^5.

Teacher hands out to each student the first 3 pages of the Practical Worksheet for students to fill in Stage I: Understand the Problem and Stage 2: Devise a Plan. They are then to write their solution in Section 3: Carry out the Plan.

Teacher walks around and suggests heuristics, especially for a plan.

Discussion of POD (10 min.)

Ask students for their answers. Ask a student with the correct answer and the correct reasoning to explain his problem solving process.

Finetune student's solution so as to highlight the use of the heuristic *Use suitable numbers (instead of algebra)* to build confidence in the veracity of the statement. Highlight the use of the heuristic *Divide into cases* according to the last digit of n as a plan to solve the problem. (Write the name of the problem Write the name of the problem beside the relevant heuristics on the notice board.)

Closure (3 min.)

Ask students for the meaning of the word *Heuristics*. Ask which heuristics were used by them in solving the POD. Ask what *Resources* they would need for the POD.

Homework: Tying a shoelace

A man is in a hurry to get on a plane. While walking as fast as he can in the airport, he notices that the lace on one of his shoes is untied. The untied shoelace will not slow him down but it may trip him up, so he must arrive at the embarkation gate with the lace tied. There are travelators (moving horizontal pedestrian carriers) in some sections of the airport along his way. Should he tie his lace on 'stationary' ground or on the travelator?

Lesson 5: The Mathematics Practical

Review of Homework (7 min.)
Ask students for their answers. Finetune students' solutions so as to highlight *Use suitable numbers (instead of algebra)* such as speed of walking = 2m/s, speed of travelator = 1m/s, distance on ground = 60m, distance on travelator = 60m, and time to tie shoelace = 3s for an initial plan. By working on different scenarios, show that it is better to tie the lace on 'stationary' ground (the later realization being that it will maximize the time on the travelator where the relative speed of 3m/s). Finally, solve the general problem using algebra. (Write the name of the problem beside the relevant heuristics on the notice board.)

Discussion: The Practical Paradigm (8 min.)
As we look at the nature of science we see two quite distinct strands: the knowledge, the important content and concepts of science and their interrelationships, and also the processes which a scientist uses in his working life. In teaching science we should be concerned both with introducing students to the important body of scientific knowledge, that they might understand and enjoy it, and also with familiarizing students with the way a problem-solving scientist works, that they too might develop such habits and use them in their own lives.
~ Woolnough and Allsop

Use the quote above and replace 'science' with 'mathematics' (you may do a 'Replace' using MS Word). Explain that the science practical lesson is very much accepted by students as part of science education and many have an understanding that it is to teach them how to 'do' science the way scientists do it. The teacher expresses his hope that similarly students will understand that we need a special lesson to learn how to 'do' mathematics the way mathematicians do it. This paradigm shift is intended to change the way students look at these 'difficult' problems which had to be done in this 'special' class.

The students have had some practice working on the first 3 pages of the Practical Worksheet. Now, introduce the complete practical worksheet as an integral part of the Mathematics Practical as follows:

- The worksheet consists of 4 pages, each page corresponding to one Pólya stage. Ideally, the student will follow the model and go through all 4 stages, with suitable loopbacks (work on blank piece of paper and attach at the relevant place).

- However, so as not to straitjacket an unwilling problem solver, the student may have a 'free' attempt and jump straight to Stage 3: Carry out the Plan and is given up to 15 min.utes to solve the problem there. A student who completes the problem in time needs only do the last Pólya stage. If not, it is explained to him that the Pólya model would be better than what he has achieved and so he is required to go through all four Pólya stages.

- The choice allowed here is to show that explicit use of Pólya's model is not necessary all the time but would be very useful when one is stuck.

Problem Solving: Teacher models (37 min.)

<u>Problem printed on Practical Worksheet to be given to students: The Lockers Problem</u>

The new school has exactly 343 lockers numbered 1 to 343, and exactly 343 students. On the first day of school, the students meet outside the building and agree on the following plan. The first student will enter the school and open all the lockers. The second student will then enter the school and close every locker with an even number. The third student will then 'reverse' every third locker; i.e. if the locker is closed, he will open it, and if the locker is open, he will close it. The fourth student will reverse every fourth locker, and so on until all 343 students in turn have entered the building and reversed the relevant lockers. Which lockers will finally remain open?

- Teacher informs students that he will work together with the class on this problem using the Practical Worksheet. Each student will be given a practical worksheet.
- To save time, the first 15 minutes of 'free' attempt will not be given.
- Teacher goes through the questions of Stage 1 and asks the students for their responses. For the purpose of modeling the process, he informs the class that he will write down the 'responses' from the point of view of himself as a student. The students on the other hand are encouraged to write down their own responses in their own practical worksheet.
- At each stage, the teacher invites class discussion and responses, but finally writes the 'responses' on the board from the point of view of himself as a student. These responses should incorporate student responses where appropriate.
- The teacher may use the earlier discussion of the Lockers Problem as a guide. The 'model answer' practical worksheet (certainly in less detail) should be prepared earlier and given to each student at the end of the lesson.
- Remember to tick off the relevant heuristics on the notice board: *Act it out, Consider a simpler problem (smaller numbers), Look for patterns, Use a suitable representation, Restate the problem in another way.*

Closure (3 min.)

Ask students for their initial impression of the practical worksheet. Remin.d the students that the worksheet is meant to help them by scaffolding Pólya's problem solving strategy. Ask the students why science has practical lessons. Ask them to reflect on the analogous idea of a mathematics practical. Tell them to do their homework on a practical worksheet.

Homework: Number of Squares

The figure is a 7×7 array where each cell is a square. Find the number of squares contained in this 7×7 array.

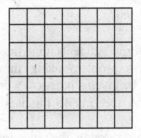

Lesson 6: Check and Expand

Review of Homework (7 min.)

Teacher models and solves the problem on the board. Use practical worksheet as scaffolding. Have some discussion to understand what a square is. Students may be given time to rethink that neither 49 nor 50 is the answer. Use the heuristic *Draw a diagram* to show that the squares to be counted are of various sizes. Use the heuristics *Consider a simpler problem (smaller numbers)* and *Divide into cases* to devise a plan: work out the solution for the 4×4 problem by considering squares of different sizes. Use the heuristic *Look for patterns* to devise a second plan to build from the simpler problem to a solution for the 7×7 problem. (Write the name of the problem beside the relevant heuristics on the notice board.) State that Stage 4 will be elaborated in the next part of the lesson.

Write the name of the problem

Discussion: Check and Expand (10 min.)

The teacher explains that this stage is called 'Looking Back' in Pólya's original model. The idea is that we do not stop when the problem is 'solved' but we should look back at the solution to check and to see what we can learn from it. We have decided to call it 'Check and Expand' to make the following features of looking back and looking forward clearer:

o   check the solution
o   find out if the method can be used to solve other problems
o   pose new problems from the original problem
o   suggest alternative solutions

The teacher then focuses on 'new problems from old' and outlines the three facets of adapting, extending and generalizing together with examples based on the Number of Squares problem:

Adapt: To change certain features of the problem.
Eg.1:   Change some numbers.
Change to a 50×50 array.

Eg.2:   Change some conditions.
Have a gap in a corner of the array.

Eg.3:   Consider the converse.
If there are 140 squares, what is the size of the array?

<u>Extend:</u>   To consider problems which are more 'difficult' or
which have greater scope.

Eg.1:   Change to a 4 × 7 rectangular array.

Eg.2:   Change to finding the number of cubes in a 7 × 7 × 7 cube made
of 1 × 1 × 1 cubes.

<u>Generalize:</u>   To consider problems which would include the given
problem as a special example.

Eg.1:   Change to an $n \times n$ array.

Eg.2:   Change to an $m \times n$ rectangular array.

Eg.3:   Change to finding the number of rectangles in an $m \times n$
rectangular array.

Problem of the Day (use Practical Worksheet): Modulus Function (30 min.)

It is given that $|x| = \begin{cases} x & \text{if } x \geq 0; \\ -x & \text{otherwise.} \end{cases}$ Sketch the graph of $y = 3|x| - 4$.

Teacher walks around and suggests heuristics, especially to understand the problem.

Discussion of POD (6 min.)

Ask students for their answers. Ask a student with the correct answer and the correct reasoning to explain his problem solving process.

Finetune student's solution so as to highlight the use of the heuristics *Consider a simpler problem* $y = |x|$ and *Use suitable numbers (instead of algebra)* to understand the modulus function and also (with suitable coordinates) to get a feel for the shape of the graph, i.e. to understand the problem. *Think of a related problem*, i.e. graph of $y = 3x - 4$ and graphs of straight lines in general to speed up the sketch. (Write the name of the problem beside the relevant heuristics on the notice board.) Pose only one problem for Stage 4:

Generalize:

Sketch the graph of $y = m|x| + c$.

Closure (2 min.)

Emphasise to students that by looking back, even if nothing else is gained, they will understand the original problem much better. Tell students that this 4th Pólya stage is very important to real mathematical problem solving and real mathematical thinking.

Homework:

Complete the 4th stage in the Practical Worksheet for the Modulus Function Problem by posing as many problems as possible.

Lesson 7: More on adapting, extending and generalising

Review of Homework and Discussion (20 min.)

The teacher reiterates the idea that we do not stop when the problem is 'solved' but we should look back at the solution to check and to see what we can learn from it. Ask the class to recall the features of Check and Expand and write them on the board:

- o check the solution
- o find out if the method can be used to solve other problems
- o pose new problems from the original problem
- o suggest alternative solutions

The teacher then focuses on the 'new' problems posed by the class based on the Modulus Function Problem. Write these under the three facets of adapting, extending and generalizing. The following are some possibilities:

Adapt: To change certain features of the problem.

Eg.1:   Change some numbers.

Change to $2y = 3|x| - 4$

Eg.2:   Change some conditions.

Find the min.imum value of y if $y = 3|x| - 4$, where x is a real number.

Extend: To consider problems which are more 'difficult' or which have greater scope.

Eg.1:   Change to $y = 3|x| + m$.

Eg.2:   Change to $y = |x - 3|$.

Eg.3:   Change to $|y| = 3|x| + 4$.

<u>Generalize:</u>   To consider problems which would include the given problem as a special example.

Eg.1:   Change to $y = m|x| + c$.

Eg.2:   Change to $ay + b|x| + c = 0$.

Eg.3:   Change to $y = f(|x|)$, where $f(x)$ is a function of x.

Problem of the Day (use Practical Worksheet): Binary Representation (25 min.)

The base 2 (binary) representation of a positive integer n is the sequence $a_k a_{k-1} a_{k-2} \ldots a_1 a_0$, where
$$n = a_k 2^k + a_{k-1} 2^{k-1} + a_{k-2} 2^{k-2} + \ldots + a_1 2^1 + a_0 2^0,$$

$a_k = 1$, and $a_i = 0$ or 1 for $i = 0, 1, \ldots, k-1$.
Write down the binary representations of the day today and of the day of the month in which you were born.

Teacher walks around and suggests heuristics, especially to understand the problem.

Discussion of POD (8 min.)

Ask students for their answers. Finetune students' solutions so as to highlight the use of the heuristic *Use numbers* to get a feel for the algebra, i.e. to understand the problem. *Think of a related problem*, i.e. decimal representation to understand the problem. (Write the name of the problem beside the relevant heuristics on the notice board.) Ask students for new problems. Pose the following problem and discuss it.

Adapt:
How would an octopus represent what we as humans normally write as 49?

Closure (2 min.)
Emphasise to students that by looking back, even if nothing else is gained, they will understand the original problem much better. Tell students that this 4th Pólya stage is very important to real mathematical problem solving and real mathematical thinking.

Homework (Weights Problem):
Weights can be placed on the left pan of a standard two-pan balance to weigh gold which is placed on the right pan. Suppose we want to be able to weigh gold in any positive integer up to 100 grams. Show that having 7 weights will be enough.

Lesson 8: Schoenfeld's Framework

Discussion: Schoenfeld's Framework (10 min.)

Explain to the students that Alan Schoenfeld was a mathematics professor who discovered Pólya's model only after he obtained his PhD. Schoenfeld grappled with the apparent worth of Pólya's model and the real-world failure of its application in the classroom. His research led him to realize that there was more than just a direct application of the model and that in fact, other factors are crucial in successful problem solving. His research culmin.ates in the construction of a framework for the analysis of complex problem-solving behaviour. The framework, in his words, consists of four aspects:

1. Cognitive resources – the body of facts and procedures at one's disposal.
2. Heuristics – 'rules of thumb' for making progress in difficult situations.
3. Control – having to do with the efficiency with which individuals utilise the knowledge at their disposal. Sometimes, this is referred to as metacognition, which can be roughly translated as 'thinking about one's own thinking'.
4. Belief systems – one's perspectives regarding the nature of a discipline and how one goes about working on it.

With regard to our attempts in problem solving, Schoenfeld's framework makes us realize that Pólya's model or heuristics is not all there is. It allows the problem solver to have a frame through which he can focus when difficulties arise.

Specifically elaborate on Control – put the following on the board.

1. These are questions to ask oneself to monitor one's thinking.
* What (exactly) am I doing? [Describe it precisely.]
  Be clear what I am doing NOW.
* Why am I doing it? [Tell how it fits into the solution.]

Be clear what I am doing in the context of the BIG picture – the
solution. Be clear what I am going to do NEXT.

2.   Stop and reassess your options when you

*   cannot answer the questions satisfactorily [probably you are on the
    wrong track]; OR
*   are stuck in what you are doing [the track may not be right or it is
    right but it is at that moment too difficult for you].

3.   Decide if you want to

*   carry on with the plan,
*   abandon the plan, OR
*   put on hold and try another plan.

Review of Homework (7 min.)

Teacher models and solves the problem on the board. Highlight Control
questions when solving the problem. Use the heuristics *Act it out* and
Use numbers to work out how to weigh 1, 2, 3, 4, 20 grams. Check if
students understand the problem, especially to show that 7 weights is
enough. Ask students what set of 7 weights they have. Use the heuristic
Consider a simpler problem (smaller numbers) to work out the
min.imum for 1, 2, 3, 4, 5, 6, 7, 8 grams. Show that the weights 1, 2, 4
will suffice up to 7 grams but 8 will be needed for 8 grams. *Looking for
patterns* will suggest powers of 2. In fact, powers of 2 from 0 to 6 will
suffice. Show that this is linked to the binary representation of a number.
Generalize the problem to weights up to n grams. (We will need x
weights where $2^{x-1} \leq n < 2^x$.)

Problem of the Day (use Practical Worksheet): Averages (27 min.)

a)   A boy claims that when he left school X and joined school Y, he
       raised the average IQ of both schools. Explain if this is possible.
b)   A striker in football is rated according to the average number of
       goals he scores in a game. Wayne had a higher average than

Carlos for games in the year 2007. He also had a higher average than Carlos for games in the year 2008. Can we say that Wayne is a better striker than Carlos over the years 2007-2008?

Teacher explains that this time, he will be walking around and asking the two control questions. Students are to respond to them appropriately otherwise to acknowledge that they are probably on the wrong track or 'stuck'.

Discussion of POD (9 min.)

Ask students for their answers. For (a) and for (b), ask a student each with the correct answer and the correct reasoning to explain his problem solving process.

Finetune student's solution so as to highlight the heuristic *Use suitable numbers (instead of algebra)* and *Guess-and-Check* to understand the problem and for a plan. Highlight the Control episodes when the teacher makes decisions on his problem solving. Example of Control episode: After a few examples of various IQs for the boy and the schools, the teacher asks: Why am I doing this? He answers that he is trying to find a positive example which will solve the problem. He then asks if he should persevere. He answers that he should persevere but be more systematic in his choice of numbers, such as whether the boy's IQ is above or below the original school and whether the original school's average IQ is above or below the new school. For (b), we assume that the 'obvious' answer that Wayne is always better can be questioned. Thus, we *work backwards* by having total scores such that Carlos is better than Wayne over two years and then trying to break those total scores into two years such that Wayne is better in each of the two years. Highlight the need for the knowledge (Resources) that an example will suffice for (a) and a counter-example will suffice for (b).

Closure (2 min.)

Ask students for the 4 components of Schoenfeld's framework. Ask students for the 2 questions to be asked for Control. Encourage students to make Control explicit when working on their Practical Worksheets, there being a column in Stage 3 for that.

Homework (use Practical Worksheet): Intersection of Two Squares

Two squares, each s on a side, are placed such that the corner of one square lies on the centre of the other. Describe, in terms of s, the range of possible areas representing the intersections of the two squares.

Lesson 9: More on Control

Review of Homework (8 min.)

Teacher models and solves the problem on the board. Highlight the Control episodes when the teacher makes decisions on his problem solving. *Consider a simpler problem (special case – tighten conditions)* for the special cases when the sides of the 'rotating' square pass through the vertices of the 'stationary' square and when they pass through the midpoint of the edges of the 'stationary' square will both result in the area $\frac{s^2}{4}$. Make a decision to pursue the conjecture that all areas are $\frac{s^2}{4}$. *Draw a diagram* of a general orientation of the two squares. Persevere and *relate to a similar problem* (i.e. one of the two special cases) to see how the area of the special case transforms to the area of the general case. Use congruent triangles to prove that the area is indeed $\frac{s^2}{4}$. (Write the name of the problem beside the relevant heuristics on the notice board.)

Discussion: Control (5 min.)

Reiterate the following on Control:

1.   These are questions to ask oneself to monitor one's thinking.
   •   What (exactly) am I doing? [Describe it precisely.]
        Be clear what I am doing NOW.
   •   Why am I doing it? [Tell how it fits into the solution.]
        Be clear what I am doing in the context of the BIG picture – the solution. Be clear what I am going to do NEXT.

2.   Stop and reassess your options when you
   •   cannot answer the questions satisfactorily [probably you are on the wrong track]; OR
   •   are stuck in what you are doing [the track may not be right or it is right but it is at that moment too difficult for you].

3.   Decide if you want to
   •   carry on with the plan,
   •   abandon the plan, OR
   •   put on hold and try another plan.

Highlight the need in (3) to make a decision to move out of the 'stuck' situation.

Problem of the Day (use Practical Worksheet): Regions in a Circle (20 min.)

A circle has n points on its circumference. All possible chords are drawn and no three chords are concurrent. Find the number of regions that the chords divide the circle into for $n = 1, 2, 3, 4, 5, 6$.

Teacher walks around and asks Control questions.

Discussion of POD (20 min.)

Ask students for their answers. Ask a student with the correct answer and the correct reasoning to explain his problem solving process.

Finetune students' solutions so as to highlight the Control episodes when the teacher makes decisions on his problem solving. *Draw a large diagram* neatly and show that the answers are 1, 2, 4, 8, 16, 31.

Discuss 'fixation' – a preoccupation with one subject, issue, etc.; obsession (from dictionary.com). Show how fixation on the pattern 1, 2, 4, 8, 16 blinds some problem solvers from the earlier statement that 'pattern is not proof'. Show a further example: Write the polynomial $x^6 - 21x^5 + 175x^4 - 735x^3 + 1624x^2 - 1764x + 723$ on the board (the polynomial is actually the expansion of $(x - 1)(x - 2)...(x - 6) + 3$). Ask the students to use their calculators to find the values of the polynomial when $x = 1, 2, 3$ and 4. Ask the students to make conjectures about the polynomial. Again if some are fixated on the supposed 'pattern', they will conjecture that the polynomial is always 3. Show how the polynomial is obtained from the 'factorised' form which leads to the value 3 only for values 1 through 6. Suggest to students that they should be more doubting of a 'pattern' when they cannot find a good reason why a pattern should continue in a particular way. Make them aware that 'fixation' either in patterns or a solution plan will mean loss of control and be a barrier to successful problem solving.

Closure (2 min.)
Ask students for the 2 questions to be asked for Control. Ask students what decision they have to make when they realize that they are 'stuck'.

Homework (use Practical Worksheet) Nice Numbers I:
A 'nice' number is a number that can be expressed as the sum of a string of two or more consecutive positive integers. Determine which of the numbers from 50 to 70 inclusive are nice.

Lesson 10: Revision

Review of Homework (8 min.)

Teacher models and solves the problem on the board. *Solve a simpler problem (smaller numbers)* and *look for patterns* to conjecture that the 'not nice' numbers are powers of 2. Use brute force to show that 64 is not 'nice'. Ask the students to help you show that the rest are 'nice' by producing suitable sums. *Use equations/algebra*, in particular, $n = a + (a + 1) + \ldots + (a + r - 1) = \frac{r(2a+r-1)}{2}$ for a plan to work on. (Write the name of the problem beside the relevant heuristics on the notice board.)

Discussion: Review of Module (15 min.)

Discuss with the class the following questions.

- What is a problem?
- Draw the model of Pólya's problem solving strategy.
- What are heuristics? Name some of them.
- Name the 4 components of Schoenfeld's framework.
- How do you know if you are 'stuck'?
- What do you do when you are 'stuck'?

Problem of the Day (use Practical Worksheet): Nice numbers II (24 min.)

n is a positive integer that is not a power of 2. Show that n is a nice number.

Teacher walks around and asks Control questions.

Discussion of POD (7 min.)

Ask students for their answers. Ask a student with the correct answer and the correct reasoning to explain his problem solving process.

Finetune student's solutions so as to highlight the Control episodes when the teacher makes decisions on his problem solving. *Think of a related problem* (Nice numbers I) to decide on an algorithm to get a sum of a string of two or more consecutive positive integers for *n*. *Divide into cases* to work on different sets of integers. *Use equations/algebra* to write out an algorithm.

Closure (1 min.)

Remin.d students of Practical Test in the next lesson.

Chapter 4

Scaffolding Suggestions, Solutions to the Problems and Assessment Notes

4.1 Introduction

This chapter provides all the 17 questions that are used in the Mathematics Practical lessons. The detailed solution of each problem is given, together with the suggested scaffoldings that teachers should use with their students in the Practical lessons. Assessment notes associated with each problem is also provided.

Problem 1: Jugs

Heuristics to highlight:
Draw a diagram; Act it out; Guess-and-check; Use equations/algebra.

You are given two jugs, one holds 5 litres of water when full and the other holds 3 litres of water when full. There are no markings on either jug and the cross-section of each jug is not uniform. Show how to measure out exactly 4 litres of water from a fountain.
Show also the following:

   i.   Get 2 litres from 3 litre and 7 litre jugs.

   ii.   Get 6 litres from 12 litre and 16 litre jugs.

  iii.   Get 12 litres from 18 litre and 24 litre jugs.

Suitable hints for Polya Stages I, II and IV

I   **Understand the problem**

(c)   Write down the heuristics you used to understand the problem.

> *Act it out and Draw a diagram – of the water level in the jugs at each step.*
> *Restate the problem in another way – Jug must be completely empty before it is filled and each time water is poured into a jug, the jug must be filled to the rim.*

II   **Devise a plan**

(a)   Write down the key concepts that might be involved in solving the problem.
     Divisibility.

(c)   Write out each plan concisely and clearly.

Plan 1   *Aim for sub-goals – work on each part of the problem one after the other.*
Act it out and Draw a diagram – of the water level in the jugs at each step until 4 litres is obtained.

Plan 2   *Use equations/algebra for (ii) – express the number of times one jug is filled as X and the number of times the other jug is emptied as Y.*

IV   Check and Expand
(a)   Write down how you checked your solution.
By explaining the procedure to a group member.
By trying other values for the jugs and the amount required.

(b)   Write down a sketch of any alternative solution(s) that you can think of.

Work until a factor of the amount required is obtained (for example 2l out of the required 4l). Then repeat the procedure the necessary number of times to obtain the required amount (in the example, twice).

(c)   Give at least one adaptation, extension or generalisation of the problem.

Adaptation:
You are given two jugs, one holds 5 litres of water when full and the other holds 3 litres of water when full. The cross-section of each jug is not uniform. There are no markings on the 5 litre jug but the halfway mark on the 3 litre jug is indicated. Show how to measure out exactly 3.5 litres of water from a fountain.

Generalisation:
You are given two jugs, one holds m litres of water when full and the other holds n litres of water when full. The cross-section of each jug is not uniform and there are no markings on both jugs. Show how to measure out exactly k litres of water from a fountain. What amounts of water cannot be measured out?

Extension:
You are given three jugs, holding a, b, and c litres of water when
full respectively. The cross-section of each jug is not uniform and
there are no markings on any of them. Show how to measure out
exactly k litres of water from a fountain. What amounts of water
cannot be measured out?

Solution(s)

Let the number of times the $5l$ jug is filled be x and the number of times
the $3l$ jug is filled be y. Note that x and y must be integers since there
are no markings on the jugs. Note also that negative values for x or y
implies 'emptied' $|x|$ times.

Thus, $5x + 3y = 4$. By inspection, $x = -1$ and $y = 3$ is a solution. (Note
that there are infinite number of solutions.) Hence the $3l$ jug must be
filled 3 times and the $5l$ jug must be poured out 1 time. Explicitly: fill
the $3l$ jug, pour into $5l$ jug, fill $3l$ jug (2nd time), pour $2l$ into $5l$ jug,
empty $5l$ jug, pour $1l$ into $5l$ jug, fill $3l$ jug (3rd time), pour into $5l$ jug to
obtain $4l$.

(i)   Get 2 litres from 3 litre and 7 litre jugs.
Use the equation $3x + 7y = 2$. By inspection, $x = 3$ and $y = -1$ is a
solution. Hence the $3l$ jug must be filled 3 times and the $7l$ jug must be
poured out 1 time. Explicitly: fill the $3l$ jug, pour into $7l$ jug, fill $3l$ jug
(2nd time), pour into $7l$ jug, fill $3l$ jug (3rd time), pour $1l$ into $7l$ jug to
obtain $2l$, (empty $7l$ jug).

(ii)   Get 6 litres from 12 litre and 16 litre jugs.
Use the equation $12x + 16y = 6$. Dividing by 4 throughout, we obtain
$3x + 4y = \frac{3}{2}$. Since the right hand side is not an integer, there are no
integer solutions for x and y. Thus, the required 6 litres cannot be
obtained.

(iii)   Get 12 litres from 18 litre and 24 litre jugs.
Use the equation $18x + 24y = 12$. Dividing by 6 throughout, we obtain
$3x + 4y = 2$. (Note the connection between the problem and getting

2 litres from 3 litre and 4 litre jugs.) By inspection, $x = 2$ and $y = -1$ is a solution. Hence the $18l$ jug must be filled 2 times and the $24l$ jug must be poured out 1 time. Explicitly: fill the $18l$ jug, pour into $24l$ jug, fill $18l$ jug (2nd time), pour $6l$ into $24l$ jug to obtain $12l$, (empty $24l$ jug).

Possible student responses

Happy to be able to solve the initial part – praise and encourage the student.

Stuck at (ii) – ask if they think it is possible. If it is not, can they try to prove their assertion?

Assessment notes

Any systematic word or diagrammatic explanation for the possible situations that is clear is acceptable. A mathematical explanation involving divisibility is needed for (ii) to be accepted. If (ii) is not acceptably solved, mark the solution as partially correct.

Problem 2: Last Digit

Heuristics to highlight:
Consider a simpler problem (smaller numbers); Make a systematic list; Look for patterns.

Find the last digit of 13^{77}.

Suitable hints for Pólya Stages I, II and IV

I   Understand the problem
(c)   Write down the heuristics you used to understand the problem.
        Consider a simpler problem (smaller numbers) – find the last digit for 13^1, 13^2, ..., 13^5.

II   Devise a plan
(a)   Write down the key concepts that might be involved in solving the problem.

        Multiplication.

(c)   Write out each plan concisely and clearly.

Plan 1   *Consider a simpler problem (smaller numbers), Make a systematic list, Look for patterns – make a table for the last digits of 13^1, 13^2, ... until a pattern is observed.*

Plan 2   *Restate the problem in another way – Starting with 13 as the first step, at each subsequent step multiply by 13 and take the last digit of the result. What is the final result after 77 steps?*
                – Find the period of a cycle (i.e. how many numbers a cycle contains). Then work out which number 77 corresponds to in the cycle.

IV   **Check and Expand**

(a)   Write down how you checked your solution.

Check with the calculator some small powers of 13 to confirm that the cycle approach works.

(b)   Write down a sketch of any alternative solution(s) that you can think of.

Expressing the solution algebraically
Suppose $x = 10a + b$, where b is the last digit of x. Then $13x = 13(10a + b) = 10(13a + b) + 3b$. Thus the last digit of $13x$ is the last digit of $3b$. This shows that only the product of the last digit of each number needs to be considered at each stage.
Hence, starting with 3 as the first step, at each subsequent step multiply by 3 and take the last digit of the result. What is the final result after 77 steps?

(c)   Give at least one adaptation, extension or generalisation of the problem.

Adaptation:
What is the greatest positive integer n such that the last digit of 13^n is 7 and 13^n is not more than 10^{99}?
Is there an integer n such that $12345123451234512345 = 13^n$?

Generalisation:
Find the last digit of 13^n, where n is a positive integer.

Extension:
Find the last two digits of 13^{77}.
Find the last digit of $1962^{2009} + 2009^{1962}$.
Find the last digit of $1962^{2009} \times 2009^{1962}$.

Solution(s)

Let the last digit of a number n be n^*.

Suppose $x = 10a + x^*$. Then $13x = 13(10a + x^*) = 10(13a + x^*) + 3x^*$. Thus $(13x)^* = (3x^*)^*$. This shows that only the product of the last digit of each number needs to be considered at each stage.

$(3^{4n+r})^* = ((3^4)^n \times 3^r)^* = (81^n \times 3^r)^* = (1^n \times 3^r)^* = (1 \times 3^r)^* = (3^r)^*$. Thus, the last digits are in a 4-cycle, i.e. 3, 9, 7, 1. Since $77 = 19 \times 4 + 1$, the last digit of 13^{77} is 3.

Possible student responses

Use the calculator – explain that the calculator can only go as far as the number of digits of the display.

Obtain the pattern from simpler problems and immediately extend the solution to 3^{77} without providing the reason for it – explain that 'pattern is NOT proof'.

Assessment notes

Obtaining the pattern from simpler problems and immediately extending the solution to the 3^{77} without providing the reason for it is only a partially correct solution.

Any systematic word or diagrammatic explanation for the periodic behavior of the last digit that is clear is acceptable.

Problem 3: Last Digit II

Heuristics to highlight
Think of a related problem; Aim for subgoals.

Find the last digit of $1962^{2009} + 2009^{1962}$.

Suitable hints for Pólya Stages I, II and IV

I   Understand the problem
(c)   Write down the heuristics you used to understand the problem.

Think of a related problem – find the last digit for 13^{77}.

II   Devise a plan
(a)   Write down the key concepts that might be involved in solving the problem.

Multiplication and Addition.

(c)   Write out each plan concisely and clearly.

Aim for subgoals – find the last digit of 1962^{2009}, find the last digit of 2009^{1962}, and finally add them up.

IV   Check and Expand
(a)   Write down how you checked your solution.

Check with the calculator some small powers of 2 and some small powers of 9 to confirm the periods of their respective cycles.

(b)   Write down a sketch of any alternative solution(s) that you can think of.

Find the last digit of $1962^{1962} + 2009^{1962}$ by finding the period of the combined cycle. Then add the last digit of 1962^{47}. (This method is not as nice as the first but gives an idea for new problems.)

(c)   Give at least one adaptation, extension or generalisation of the problem.

Adaptation:
What is the least positive integer n such that the last digit of $1962^{n} + 2009^{1962}$ is 7?

Generalisation:
Find the last digit of $1962^{m} + 2009^{n}$, where m and n are positive integers.

Extension:
Find the last two digits of $1962^{2009} + 2009^{1962}$.

Solution(s)

We have shown before that only the product of the last digit of each number needs to be considered at each stage, and that the last digits are periodic.

The last digits starting with 2 are in the sequence 2, 4, 8, 6, 2, 4, 8, 6, ... with a 4-period. $2009 \div 4$ leaves a remainder of 1. Thus, the last digit of 1962^{2009} is the same as the last digit of 2^1, i.e. 2.
The last digits starting with 9 are in the sequence 9, 1, 9, 1, ... with a 2-period. $1962 \div 2$ leaves a remainder of 0. Thus, the last digit of 2009^{1962} is the same as the last digit of 9^2, i.e. 1.

Hence, the last digit of $1962^{2009} + 2009^{1962}$ is $2 + 1 = 3$.

Possible student responses

Use the calculator – explain that the calculator can only go as far as the number of digits of the display.

Obtain the pattern from simpler problems and immediately extend the solution to $1962^{2009} + 2009^{1962}$ without providing the reason for it – explain that 'pattern is NOT proof'.

Assessment notes

Obtaining the pattern from simpler problems and immediately extending the solution to the $1962^{2009} + 2009^{1962}$ without providing the reason for it is only a partially correct solution.

Any systematic word or diagrammatic explanation for the periodic behavior of the last digit that is clear is acceptable. Stating that 'only the product of the last digit of each number needs to be considered at each stage, and that the last digits are periodic' based on a previous result is acceptable.

Problem 4: Phoney Russian Roulette

Heuristics to highlight:
Act it out; Draw a diagram; Restate the problem in another way; Aim for sub-goals.

Two bullets are placed in two consecutive chambers of a 6-chamber revolver. The cylinder is then spun. Two persons play a safe version of Russian Roulette. The first points the gun at his hand phone and pulls the trigger. The shot is blank. Suppose you are the second person and it is now your turn to point the gun at your hand phone and pull the trigger. Should you pull the trigger or spin. the cylinder another time before pulling the trigger?

Suitable hints for Pólya Stages I, II and IV

I   Understand the problem
(c)   Write down the heuristics you used to understand the problem.

> *Act it out and Draw a diagram – of the 6 chambers in the revolver and two consecutive loaded chambers.*
> *Restate the problem in another way – Which is less likely to destroy your phone?*

II   Devise a plan
(a)   Write down the key concepts that might be involved in solving the problem.

> *Probability, sample space.*

(c)   Write out each plan concisely and clearly.

Plan 1   *Aim for sub-goals – calculate the probability of destroying the phone if the cylinder is spun again, and the probability if the cylinder is not spun.*

Plan 2   *Aim for subgoals – to find the probability if the cylinder is not spun, first find the sample space, i.e. the list of chambers that can be hit by the hammer of the gun.*

IV   Check and Expand
(a)   Write down how you checked your solution.

By explaining the solution to a group member.
By setting up an experiment to simulate the problem and trying out a suitable number of iterations. A simple paper disc can be made and spun.

(b)   Write down a sketch of any alternative solution(s) that you can think of.

Generalise the problem to m consecutive bullets in n chambers to obtain a general solution. Then see if the solution holds true for m = 2 and n = 6.

(c)   Give at least one adaptation, extension or generalisation of the problem.

Adaptation 1:
Two bullets are placed in two consecutive chambers of a 6-chamber revolver. Two persons play a safe version of Russian Roulette. The first points the gun at his hand phone and pulls the trigger. The gun fires and the first hand phone is destroyed. Suppose you are the second person and it is now your turn to point the gun at your hand phone and pull the trigger. Should you pull the trigger or spin the cylinder another time before pulling the trigger?

Adaptation 2:
Two bullets are placed in two chambers of a 6-chamber revolver
such that there is exactly one empty chamber between them. The
cylinder is then spun. Two persons play a safe version of Russian
Roulette. The first points the gun at his hand phone and pulls the
trigger. The shot is blank. Suppose you are the second person
and it is now your turn to point the gun at your hand phone and
pull the trigger. Should you pull the trigger or spin the cylinder
another time before pulling the trigger

Generalisation:
m bullets are placed in consecutive chambers of an n-chamber.
The cylinder is then spun. Two persons play a safe version of
Russian Roulette. The first points the gun at his hand phone and
pulls the trigger. The shot is blank. Suppose you are the second
person and it is now your turn to point the gun at your hand
phone and pull the trigger. Should you pull the trigger or spin
the cylinder another time before pulling the trigger?

Extension:
Two bullets are placed in consecutive chambers of a 6-chamber
revolver. The cylinder is then spun. Two persons play a safe
version of Russian Roulette. The first points the gun at his hand
phone and pulls the trigger. The shot is blank. The second person
does not spin the cylinder another time before pulling the
trigger. The shot is also blank. Now it is the turn of the first
person again. Should he pull the trigger or spin the cylinder
another time before pulling the trigger?

Solution(s)

Let the chambers be numbered from 1 to 6 and let the chamber spin
such that 1 is followed by 2 followed by 3 and so on until 6 is followed
by 1. We may assume that the bullets are placed in chambers 1 and 2. If
the first person gets a blank, then he would have triggered either
chambers 3, 4, 5 or 6. If you, the second person, do not spin, then you

would have the following chambers as possibilities: 4, 5, 6 or 1. There are thus only 4 outcomes in the sample space, and only one (chamber 1) is bad. Thus, the probability of destroying your phone is ¼. Since the probability of destroying your phone if you spin is clearly 2/6 = 1/3, it is better not to spin the cylinder.

Possible student responses

There is one less blank chamber if one does not spin, so it is better to spin – ask student not to be too quick to conclude but instead to be more cautious when dealing with problems (control).

Stuck at working out probability if one does not spin – ask student to concentrate on being clear about the sample space.

Assessment notes

Any systematic word or diagrammatic explanation for the sample space that is clear is acceptable. A mathematical explanation involving probability is needed to be considered correct.

Problem 5: 4-D coincidence

Heuristics to highlight
Act it out; Draw a diagram; Restate the problem in another way; Aim for sub-goals.

'*1 in 100 million chance: 6904 wins top two prizes in 4-D draw*' was the title of a news report on page H1 of The Straits Times Thursday 28 June 2007. Is this probability correct?
(The 4-D is a popular Singaporean gambling game which gives 23 prizes for 4-digit sequences obtained from 0000 to 9999. Repeats are allowed.)

Suitable hints for Pólya Stages I, II and IV

I   Understand the problem
(c)   Write down the heuristics you used to understand the problem.

Attempt 1   *Act it out – list out some possibilities for first prize and second prize.*

Attempt 2   *Act it out – would 1234 winning top two prizes also elicit the same news report? how about 7583? or any other number?*
Restate the problem in another way – what is the probability that a 4-digit sequence wins both the first and second prize?

II   Devise a plan
(a)   Write down the key concepts that might be involved in solving the problem.

Concepts involving probability and sample space.

(c)   Write out each plan concisely and clearly.

<u>Plan 1</u>   *Consider a simpler problem (smaller numbers) – calculate the probability of getting double 1 in two throws of a die by considering the sample space and the favourable outcomes. Then calculate the probability of getting 6904 in two picks from 10,000 possible 4-digit sequences.*

<u>Plan 2</u>   *Restate the problem in another way – what is the probability that a 4-digit sequence wins both the first and second prize?*

IV   Check and Expand
(a)   Write down how you checked your solution.

By setting up an experiment using two dice to simulate the problem – indicate what similar event is 'newsworthy', and then estimate the probability of that event by repeating the experiment a suitable number of times.

(b)   Write down a sketch of any alternative solution(s) that you can think of.

Find the probability that the first two prizes have different numbers. Then subtract this probability from 1.

(c)   Give at least one adaptation, extension or generalisation of the problem.

Adaptation:
What is the probability that two persons in the class have the same birthday?

Generalisation:
There are n equally likely outcomes in a sample space. What is the probability that two trials produce identical outcomes?

Extension:
There are 23 prizes in a 4-D draw. What is the probability that at least one 4-digit sequence gets at least 2 prizes?

Solution(s)

We restate the problem more clearly as: what is the probability that a 4-digit sequence wins both the first and second prize? The sample space consists of $10{,}000 \times 10{,}000$ outcomes. The favourable outcomes are $(0000, 0000)$, $(0001, 0001)$, ..., $(9999, 9999)$, a total of $10{,}000$ favourable outcomes. Thus, the required probability is $\dfrac{10000}{10000 \times 10000} = \dfrac{1}{10000}$. Hence, the newspaper report is erroneous.

Possible student responses

The probability is $\dfrac{1}{10000} \times \dfrac{1}{10000} = \dfrac{1}{10000 \times 10000}$ – ask student not to be too quick to conclude but instead to be more cautious when dealing with problems (control).

Assessment notes

Any systematic word or diagrammatic explanation for the sample space that is clear is acceptable. A mathematical explanation involving probability is needed to be considered correct.

Problem 6: Same Last Digit

Heuristics to highlight
Use suitable numbers (instead of algebra); Think of a related problem; Divide into cases.

Show that the integer n always has the same last digit as its fifth power n^5.

Suitable hints for Pólya Stages I, II and IV

I   **Understand the problem**

(c)   Write down the heuristics you used to understand the problem.

Use suitable numbers (instead of algebra) – check if it is true for $n = 1, 2, 5, 88$.

Think of a related problem – last digit.

II   **Devise a plan**

(a)   Write down the key concepts that might be involved in solving the problem.

Algebraic manipulation and factorization.

(c)   Write out each plan concisely and clearly.

Divide into cases – find the last digit of n and n^5 for $n = 0, 1, 2, ..., 9$.

IV   **Check and Expand**

(a)   Write down how you checked your solution.

Check with the calculator some values of n and n^5.

(b)   Write down a sketch of any alternative solution(s) that you can think of.

Use equations/algebra. Factorise $n^5 - n$ and show that it is divisible by 10.

(c)   Give at least one adaptation, extension or generalisation of the problem.

Adaptation:
Show that the cube of the integer n always has the same last digit as the 7th power of n.

Generalisation:
Show that the integer n always has the same last digit as n^{4k+1}, where k is a positive integer.

Extension:
Show that there is no positive integer k such that the integer n always has the same last two digits as its k-th power.

Solution(s)

Solution 1:   List out all the cases using a table. From a related problem (Last Digit), we need only consider the last digit when multiplying. We may ignore the sign of *n* and so we consider only nonnegative *n*.

Last digit of n	0	1	2	3	4	5	6	7	8	9
Last digit of n^2	0	1	4	9	6	5	6	9	4	1
Last digit of n^3	0	1	8	7	4	5	6	3	2	9

Last digit of n^4	0	1	6	1	6	5	6	1	6	1
Last digit of n^5	0	1	2	3	4	5	6	7	8	9

The table shows that the last digit of n^5 is the same as the last digit of n for all integers n.

Solution 2:   We have:
$$n^5 - n = n(n^4 - 1) = n(n^2 - 1)(n^2 + 1) = n(n - 1)(n + 1)(n^2 + 1)$$

Since one of n or $(n + 1)$ is even, when one of the four factors is a multiple of 5, the whole expression would be a multiple of 10. Thus, we need only consider the cases of the remainders when n is divided by 5. Thus, let $n = 5k + i$, where $i = 0, 1, 2, 3,$ or 4.

If $i = 0$, then n is a multiple of 5.
If $i = 1$, then $n - 1$ is a multiple of 5.
If $i = 2$, then the last digit of $n^2 + 1$ is $2^2 + 1 = 5$. Thus, $n^2 + 1$ is a multiple of 5.
If $i = 3$, then the last digit of $n^2 + 1$ is the last digit of $3^2 + 1$, i.e. 0. Thus, $n^2 + 1$ is a multiple of 5.
If $i = 4$, then $n + 1$ is a multiple of 5.

Hence, $n^5 - n$ is a multiple of 10, i.e. the last digit of n^5 is the same as the last digit of n.

Possible student responses

Students do not spend enough time trying to understand the problem before giving up – ask them to try some values for n.

Obtain the pattern from some values of n and immediately extend the solution to all n without providing the reason for it – explain that 'pattern is NOT proof'.

Assessment notes

Obtaining the pattern from a few values of n and immediately extending the solution to the all n without providing the reason for it is only a partially correct solution.

Problem 7: Tying a Shoelace

Heuristics to highlight
Use suitable numbers (instead of algebra); Use equations/algebra; Aim for sub-goals.

A man is in a hurry to get on a plane. While walking as fast as he can in the airport, he notices that the lace on one of his shoes is untied. The untied shoelace will not slow him down but he must arrive at the embarkation gate with the lace tied. There are travelators (moving horizontal pedestrian carriers) in some sections of the airport along his way. Should he tie his lace on 'stationary' ground or on the travelator?

Suitable hints for three of Pólya's stages

I   Understand the problem
(c)   Write down the heuristics you used to understand the problem.

Act it out – Observe that he can walk on the travelator.

II   Devise a plan
(a)   Write down the key concepts that might be involved in solving the problem.

Speed.

(c)   Write out each plan concisely and clearly.

Plan 1   *Use suitable numbers (instead of algebra) – let the speed of the man = speed of travelator = 1m/s; length of ground = length of travelator =10m; time taken to tie shoelace = 1s. Work out the time taken for the two cases.*

Plan 2   *Use equations/algebra – let the speed of the man = s_1 m/s, speed of the travelator = s_2 m/s, etc.*

IV   Check and Expand

(a)   Write down how you checked your solution.

           By checking with another set of suitable numbers.

(b)   Write down a sketch of any alternative solution(s) that you can think of.

           By considering the time 'saved' when tying the shoelace on the travelator.

(c)   Give at least one adaptation, extension or generalisation of the problem.

           Adaptation:
           Assuming that the walking speed of the man is 4m/s, the speed of the travelator is 2m/s, the time taken to tie his shoelace is 5s, and that he can completely tie his shoelace if he starts doing so at the beginning of any stretch of travelator, how much faster would he arrive at the embarkation gate if he should he tie his lace on the travelator rather than on 'stationary' ground?

           Generalisation:
           Suppose we do not assume that he will complete tying his shoelace within the time spent on the travelator.

           Extension:
           A man is in a hurry to get on a plane. While walking as fast as he can in the airport, he notices that the lace on one of his shoes is untied. The untied shoelace will not slow him down but he must arrive at the embarkation gate with the lace tied. There are travelators (moving horizontal pedestrian carriers) in some sections of the airport along his way and he must take a lift.

> *Should he tie his lace on 'stationary' ground, on the travelator,*
> *or in the lift?*

Solution(s)

Solution 1
Let the walking speed of the man be s_1 m/s, the speed of the travelator be s_2 m/s, the time taken to tie his shoelace is t s, the distance of 'stationary' ground is d_1 m and the distance on the travelator be d_2 m.

Suppose he ties on 'stationary' ground.
Time to tie lace = t.
Time to traverse d_1 m of 'stationary ground = $\dfrac{d_1}{s_1}$.

Time to traverse d_2 m of travelator = $\dfrac{d_2}{s_1 + s_2}$.

Thus, time taken to reach the gate = $t + \dfrac{d_1}{s_1} + \dfrac{d_2}{s_1 + s_2}$.

Suppose he ties on the travelator.
Time to traverse d_1 m of 'stationary ground = $\dfrac{d_1}{s_1}$.

Time to tie lace = t.
Distance travelled on travelator while tying lace = ts_2.

Time to traverse remaining $(d_2 - ts_2)$ m of travelator = $\dfrac{d_2 - ts_2}{s_1 + s_2}$.

Thus, time taken to reach the gate $= t + \dfrac{d_1}{s_1} + \dfrac{d_2 - ts_2}{s_1 + s_2}$ <

$t + \dfrac{d_1}{s_1} + \dfrac{d_2}{s_1 + s_2}$.

Solution 2

No matter where he ties his shoelace, he has to spend the same amount of time tying his shoelace and traversing the 'stationary' ground. If he ties on the travelator, he 'saves' the distance that he has moved on the travelator while tying. Thus because of there being less distance to traverse, the time spent traversing the travelator if he ties on the travelator is less than if he ties on 'stationary' ground.

Possible student responses

Students use their intuition to insist on any one of 3 possibilities, i.e. better on ground, better on travelator, no difference. All 3 responses have been observed in real life. Insist that student be rigourous in their proof.

Assessment notes

Only proofs that meet the rigour of the Solutions 1 or 2 are accepted as completely correct solutions.

Problem 8: Lockers

Heuristics to highlight
Act it out; Consider a simpler problem (smaller numbers); Look for patterns; Restate the problem in another way; Make a systematic list.

Pólya's stage(s) to highlight
Understand the problem; Devise a plan; Check and Expand.

The new school has exactly 343 lockers numbered 1 to 343, and exactly 343 students. On the first day of school, the students meet outside the building and agree on the following plan. The first student will enter the school and open all the lockers. The second student will then enter the school and close every locker with an even number. The third student will then 'reverse' every third locker; i.e. if the locker is closed, he will open it, and if the locker is open, he will close it. The fourth student will reverse every fourth locker, and so on until all 343 students in turn have entered the building and reversed the relevant lockers. Which lockers will finally remain open?

Suitable hints for three of Pólya's stages

I   Understand the problem
(c)   Write down the heuristics you used to understand the problem.

Attempt 1   *Act it out – Imagine you are the first few students, and act out what they will do to the first few lockers.*

<u>Attempt 2</u>   *Restate the problem in another way – What type of*
                        number causes the corresponding locker to be touched
                        an odd number of times? What feature(s) of a number
                        causes the corresponding locker to be touched? What
                        type of number has an odd number of factors?

II   Devise a plan

(a)   Write down the key concepts that might be involved in solving
         the problem.

         Factors of an integer.

(c)   Write out each plan concisely and clearly.

<u>Plan 1</u>   *Consider a simpler problem (smaller numbers), Use a*
                       suitable representation, Look for patterns – Consider
                       the first 10 lockers, and examine whether they will be
                       closed by the end when all the students have entered the
                       school; use a suitable representation to depict the
                       opening and closing of the lockers; look for patterns to
                       see which lockers remain open.

<u>Plan 2</u>   *Make a systematic list, Look for patterns – list the*
                       factors of some numbers and try to understand why a
                       square has an odd number of factors while the others
                       have even numbers.

IV   Check and Expand

(a)   Write down how you checked your solution.

         Verify for a few cases.

(b)   Write down a sketch of any alternative solution(s) that you can
         think of.

         Use prime factorisation to prove that a natural number is a
         square if and only if it has an odd number of factors.

(c)   Give at least one adaptation, extension or generalisation of the problem.

Adaptation 1:
The i-th student reverses every i-th locker except Locker i.

Adaptation 2:
The i-th student reverses every locker whose number is a factor of i.

Extension:
There are m stages to finally open the locker (for example: put the key in the lock, turn the key, pull down the handle, etc.). The i-th student goes to every i-th locker and acts out the next stage, and if the locker is fully open, closes and locks it. Which lockers remain open in the end?

Generalisation 1:
There are n lockers.

Generalisation 2:
There are n lockers. How many lockers remain open?

Solution(s)

<u>Solution 1</u>   Let n be a positive integer. Let $a_1, a_2, \ldots, a_k$ be all the k distinct factors of n with $a_1 < a_2 < \ldots < a_k$. Observe the following:

(i) n is the product of the i-th smallest factor a_i and the i-th largest factor a_{k+1-i}, i.e. $n = a_i a_{k+1-i}$, where $a_i \leq a_{k+1-i}$.

(ii)   If n is not a square, for $n = a_i a_{k+1-i}$, we have $a_i < \sqrt{n}$ and $a_{k+1-i} > \sqrt{n}$. This implies that each factor has a 'partner' factor distinct from itself such that the product of the two factors is n. Hence the number of factors must be even.

(iii)   If n is a square, then $\sqrt{n}$ is an integer. As in (ii), each factor less than $\sqrt{n}$ has a 'partner' factor distinct from itself such that the product of the two factors is n. Observe that $n = \sqrt{n} \times \sqrt{n}$ and so this factor $\sqrt{n}$ has no distinct 'partner'. Hence the number of factors must be odd.

Now each locker will be 'touched' by a student whose number is a factor of the number of the locker. A square will be touched by an odd number of students because it has an odd number of factors as shown above. In the sequence of 'open' followed by 'close', an odd number of actions will end in the locker eventually open. A non-square will by the result above be closed at the end.

Hence, the lockers that will remain open will be those whose number is a square, i.e. when there are 343 students, they are 1, 4, 9, 16, 25, 36, 49, ..., 324.

Solution 2

If a number has prime factorization $n = p_1^{a_1} p_2^{a_2} \ldots\ldots p_n^{a_n}$, then the total number of factors (the numbers 1 and itself included) is $(a_1 + 1)(a_2 + 1)(a_3 + 1)\ldots(a_n + 1)$. If this number is odd, then all the numbers $a_1, a_2, \ldots\ldots, a_n$ must be even, hence n must be a perfect square.

Possible student responses

Students may make only incomplete observations, leading to wrong generalizations. For example, prime numbered lockers are closed, hence all the other numbered lockers are open.

Assessment notes

Observing from the listing of factors that factors occur in pairs and so the only numbers that have the odd number of factors are the perfect squares is only a partially correct solution.

Any systematic word or diagrammatic explanation that the factors are divided into two groups by $\sqrt{n}$ with pairs formed from one of each set and that $\sqrt{n}$ forms a pair with itself for square n can be considered correct.

Problem 9: Number of Squares

Heuristics to highlight
Draw a diagram; Consider a simpler problem (smaller numbers); Look for patterns; Divide into cases.

The figure is a 7 × 7 array where each cell is a square. Find the number of squares contained in this 7 × 7 array.

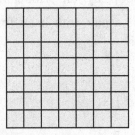

Suitable hints for three of Pólya's stages

I   Understand the problem
(c)   Write down the heuristics you used to understand the problem.

Draw a diagram – see that the squares to be counted are of various sizes.

II   Devise a plan
(a)   Write down the key concepts that might be involved in solving the problem.

Counting principles.

(c)   Write out each plan concisely and clearly.

Plan 1   *Consider a simpler problem (smaller numbers), Look for patterns – 1×1 square, 2×2 square, and 3×3 square.*

Plan 2   *Divide into cases – Classify the squares in the 7×7 square and count for each case; take the sum.*

IV   Check and Expand
(a)   Write down how you checked your solution.

        Verify the pattern of the answer for the 4×4 square by counting out all the squares 'by hand'.

(b)   Write down a sketch of any alternative solution(s) that you can think of.

        By counting the number of squares that can be formed by using each of the points as the top left vertex of the square. However, this needs some messy notation.

(c)   Give at least one adaptation, extension or generalisation of the problem.

        Adaptation:
        How many squares are there in a 7×7 square which does not contain the top-leftmost square?

        Generalisation 1:
        How many squares are there in an n×n square?

        Generalisation 2:
        How many squares are there in an m×n rectangle?

        Extension 1:
        How many cubes are there in a 7×7×7 cube? (Perhaps this problem can be solved by similarly classifying and listing.)

Extension 2:
How many rectangles are there in a 7×7 square? (A square is a special rectangle. There will be more rectangles in a 7×7 square than number of squares. However, perhaps the counting in this problem may be more complicated than the counting of squares.)

Extension 3:
How many rectangles are there in an m×n rectangle?

Solution(s)

Consider the $r \times r$, where $r = 1, 2, ..., 7$, squares in the 7×7 square. Starting with the bottom-leftmost $r \times r$ square, we can move one step to the right to get another $r \times r$ square. Continuing one step at a time to the right, we get exactly $(8 - r)$ such squares with their 'base' on the first horizontal line. Next, we move up one step and consider the leftmost $r \times r$ square with the base on the second horizontal line. As before, we move one step at a time to the right and obtain exactly $(8 - r)$ such squares with their base on the second horizontal line. We now observe that we can continue to move up one step at a time to get $(8 - r)$ squares with the base on the same horizontal line. In all, we can use only the first $(8 - r)$ horizontal lines as base lines. Thus, the number of $r \times r$ squares $= (8 - r) \times (8 - r) = (8 - r)^2$.
Thus, the number of squares in the 7×7 square $= 7^2 + 6^2 + ... + 1^2 = 140$.

Possible student responses

Count only the 1×1 squares.

Obtain the pattern from simpler problems and immediately extend the solution to the 7×7 square without providing the reason for it. (Pattern is NOT proof.)

Assessment notes

Obtaining the pattern from simpler problems and immediately extending the solution to the 7×7 square without providing the reason for it is only a partially correct solution.

Any systematic word or diagrammatic explanation that squares of a particular size can be counted 'moving' a number of steps sideways and a number of steps upwards can be considered correct.

Problem 10: Modulus Function

Heuristics to highlight
*Use suitable numbers (instead of algebra); Think of a related problem;
Divide into cases.*

It is given that $|x| = \begin{cases} x & \text{if } x \geq 0; \\ -x & \text{otherwise.} \end{cases}$ Sketch the graph of $y = 3|x| - 4$.

Suitable hints for three of Pólya's stages

I   Understand the problem
(c)   Write down the heuristics you used to understand the problem.

*Use suitable numbers (instead of algebra) – substitute some
values for x to try to interpret the piecewise function.*

II   Devise a plan
(a)   Write down the key concepts that might be involved in solving
the problem.

Curve sketching.

(c)   Write out each plan concisely and clearly.

Plan 1   *Use suitable numbers (instead of algebra) – plot some
points to get a feel for the graph.*

Plan 2   *Think of a related problem, Divide into cases - graph of
y = 3x − 4 and graphs of straight lines in general, divide
into cases according to the sign of x.*

IV   Check and Expand

(a)   Write down how you checked your solution.

By verifying if the coordinates of some points on the sketched graph fit the equation.

(b)   Write down a sketch of any alternative solution(s) that you can think of.

An algebraic approach according to the sign of x.

(c)   Give at least one adaptation, extension or generalisation of the problem.

Adaptation 1:
Sketch the graph of $y = 4|x| + 5$.

Generalisation:
Sketch the graph of $y = m|x| + c$.

Extension:
Sketch the graph of $|y| = 3|x| - 4$.

Solution(s)

Solution 1
Since $y = 3x - 4$ is a straight line, the different portions of the graph $y = 3|x| - 4$ must also be straight lines. The following are points on the curve by calculation: $(-2,2)$, $(-1,-1)$, $(0,-4)$, $(1,-1)$. These points are chosen because they represent the two portions according to the sign of x: negative and nonnegative. The two portions should be rays and meet at $(0,-4)$. We have the following graph:

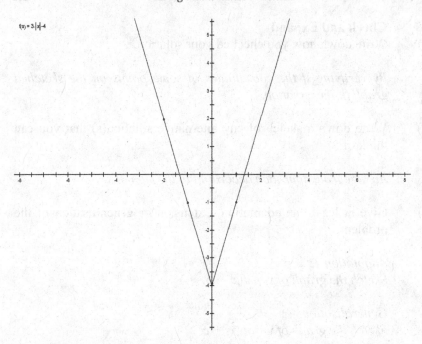

Solution 2

Let $x \geq 0$. Then $3|x| - 4 = 3x - 4$. Thus, we sketch the graph $y = 3x - 4$ for $x \geq 0$.

Let $x < 0$. Then $3|x| - 4 = -3x - 4$. Thus, we sketch the graph $y = -3x - 4$ for $x < 0$.

The resulting graph will be as above.

Possible student responses

Students complain that they have never seen such notation before – ask them to spend time trying to understand the problem, i.e. understand the new notation.

Assessment notes

Plotting of points without any analytic reasoning as outlined above is considered only as a partially correct answer.

Problem 11: Binary Representation

Heuristics to highlight
Use numbers; Consider a simpler problem (smaller numbers); Look for patterns.

The base 2 (binary) representation of a positive integer n is the sequence $a_k a_{k-1} a_{k-2} \ldots a_1 a_0$, where
$$n = a_k 2^k + a_{k-1} 2^{k-1} + a_{k-2} 2^{k-2} + \ldots + a_1 2^1 + a_0 2^0,$$
$a_k = 1$, and $a_i = 0$ or 1 for $i = 0, 1, \ldots, k-1$.
Write down the binary representations of the day today and of the day of the month in which you were born.

Suitable hints for Pólya Stages I, II and IV

I   Understand the problem
(c)   Write down the heuristics you used to understand the problem.

Aim for sub-goals – try to understand what the a_i's are.
Work backwards – start with a binary sequence and work out what n is.
Act it out, Use numbers - work out the binary representations of 1, 2, 3.

II   Devise a plan
(a)   Write down the key concepts that might be involved in solving the problem.

Indices.

(c)   Write out each plan concisely and clearly.

Consider a simpler problem (smaller numbers), Look for patterns - work out the binary representations for the first few numbers until a pattern is found. Then write down the required numbers in binary form.

IV   Check and Expand

(a)   Write down how you checked your solution.

Check with friends. Try some other numbers.

(b)   Write down a sketch of any alternative solution(s) that you can think of.

As in the definition, express the numbers as sums of powers of 2, and then put in 1's into the corresponding 'places' in the binary representation.

(c)   Give at least one adaptation, extension or generalisation of the problem.

Adaptation:
Find n if its binary representation is 2010.

Generalisation:
The base m (where m =2, 3, ..., or 10) representation of a positive integer n is the sequence $a_k a_{k-1} a_{k-2} ... a_1 a_0$, where
$$n = a_k m^k + a_{k-1} m^{k-1} + a_{k-2} m^{k-2} + ... + a_1 m^1 + a_0 m^0,$$
$a_k \neq 0$, and $a_i = 0, 1, ..., $ or $m-1$ for $i = 0, 1, ..., k$.
Write down the base 3 representation of 2010.

Extension:
A hexadecimal is a base-16 representation of a number. Find 2010 in hexadecimal form. (Note that the representation needs the letters A, B, C, D, E, and F to represent 10, 11, 12, 13, 14, and 15 respectively.)

Solution(s)

Suppose the day is 25 and the birth month is 12. Thus, we have:

$25 = 16 + 8 + 1 = 2^4 + 2^3 + 2^0 = 11001_2$.

$12 = 8 + 4 = 2^3 + 2^2 = 1100_2$.

Possible student responses

Intimidated by the notation – ask student to aim for sub-goals. Suggest working on the a_i's first.

Assessment notes

Since the notation with subscript 2 is not explicitly stated, (in base 2) is acceptable. Alternatively, statements such as 25 is 11001 is acceptable.

Problem 12: Weights

Heuristics to highlight

Act it out; Use numbers; Consider a simpler problem (smaller numbers); Look for patterns; Think of a related problem.

Weights can be placed on the left pan of a standard two-pan balance to weigh gold which is placed on the right pan. Suppose we want to be able to weigh gold in any positive integer up to 100 grams. Show that having 7 weights will be enough.

Suitable hints for Pólya Stages I, II and IV

I   **Understand the problem**

(c)   Write down the heuristics you used to understand the problem.

   Act it out, Use numbers – work out how to weigh 1, 2, 3, 4, 20 grams.

II   **Devise a plan**

(a)   Write down the key concepts that might be involved in solving the problem.

   Binary representation.

(c)   Write out each plan concisely and clearly.

Plan 1   *Consider a simpler problem (smaller numbers) – work out the minimum for 1, 2, 3, 4, 5, 6, 7, 8 grams.*
        Look for patterns - suggest powers of 2.

Plan 2   *Think of a related problem – use binary representation*
                      of numbers from 1 to 100 to show that weights 2^n, where
                      n = 0, 1, ..., 6 suffice.

IV   **Check and Expand**
(a)   Write down how you checked your solution.

              By trying other values 30 and 99.

(b)   Write down a sketch of any alternative solution(s) that you can
              think of.

              List the weights needed from 1, 2, 4, 8, 16, 32, 64 for all the
              numbers 1 to 100 using an Excel spreadsheet. It should not take
              too long.

(c)   Give at least one adaptation, extension or generalisation of the
              problem.

              Adaptation:
              You are allowed to choose 7 weights to weigh gold in any
              positive integer up to n grams. What weights would you choose
              and what is the maximum n that you can get? (You are required
              to prove that any positive integer up to n grams can be obtained
              but not required to prove that the value for n that you obtain is
              maximum.)

              Generalisation 1:
              Weights can be placed on the left pan of a standard two-pan
              balance to weigh gold which is placed on the right pan. Suppose
              we want to be able to weigh gold in any positive integer up to n
              grams. Let m be the greatest integer such that $2^m \leq n$. Show that
              having m+1 weights will be enough.

Generalisation 2:

Weights can be placed on the left pan of a standard two-pan balance to weigh gold which is placed on the right pan. Suppose we want to be able to weigh gold in any positive integer up to n grams. Let m be the greatest integer such that $2^m \leq n$. Show that having m+1 weights is the minimum number that suffices.

Extension:

A 3-pan balance is such that there are 2 left pans, one placed the same distance from the fulcrum as the right pan and the other at twice the distance from the fulcrum. Weights can be placed on the left pans to weigh gold which is placed on the right pan. Suppose we want to be able to weigh gold in any positive integer up to 100 grams. Show that having 5 weights will be enough.

Solution(s)

<u>Solution 1:</u>   Let the binary representation of a positive integer n be the sequence $a_k a_{k-1} a_{k-2} \ldots a_1 a_0$. Since $100_{10} = 1100100_2$, we need only the weights 2^m, $m = 0, 1, 2, 3, 4, 5, 6$. Thus, for any positive integer n not exceeding 100, find its binary representation and then select the weight 2^j if and only if $a_j = 1$ in its binary representation. Hence 7 weights suffice.

<u>Solution 2:</u>   List out using a table. (We will just show the first 10.)

	1	2	3	4	5	6	7	8	9	10
1	X		X		X		X		X	
2		X	X			X	X			X
4				X	X	X	X			
8								X	X	X
16										
32										
64										

Possible student responses

May miss out the need for integer weights – ask student to reread the problem, act out some weighings.

Student may have different set of 7 weights – this is possible as the solution is not unique, for e.g. replacing 64 with anything from 37 to 63 will work. However, the solution involving powers of 2 will give rise to nice extensions.

Assessment notes

Any systematic word or diagrammatic explanation that is clear is acceptable.

Problem 13: Averages

Heuristics to highlight

Use suitable numbers (instead of algebra); Guess-and-Check; Divide into cases; Work backwards.

---

a)   A boy claims that when he left school X and joined school Y, he raised the average IQ of both schools. Explain if this is possible.

b)   A striker in football is rated according to the average number of goals he scores in a game. Wayne had a higher average than Carlos for games in the year 2007. He also had a higher average than Carlos for games in the year 2008. Can we say that Wayne is a better striker than Carlos over the years 2007-2008?

Suitable hints for three of Pólya's stages

I   Understand the problem
(c)   Write down the heuristics you used to understand the problem.

Use suitable numbers (instead of algebra) – (a) Put in some values for the IQs of the boy and the schools and work out what happens to the averages when the boy changes school; (b) Put in some values for the number of goals scored by each player in each year and see how to calculate the average of each of them over 2 years.

II   Devise a plan
(a)   Write down the key concepts that might be involved in solving the problem.

Average.

(c)   Write out each plan concisely and clearly.

Plan 1   *Guess-and-Check – (a) Guess some values for the IQs of*
                  the boy and the schools and check if it fits the scenario.

Plan 2   *Divide into Cases – (a) Consider values where the IQ of*
                  the boy relative to the IQs of the school is lower or
                  higher.

Plan 3   *Work backwards – (b) Have total scores such that*
                  Carlos is better than Wayne over two years and then try
                  to break those total scores into two years such that
                  Wayne is better in each of the two years.

IV   Check and Expand
(a)   Write down how you checked your solution.

         By checking with other sets of values to make (a) true and (b)
         false.

(b)   Write down a sketch of any alternative solution(s) that you can
         think of.

         An algebraic approach for both (a) to compare the IQs after the
         boy has changed schools.

(c)   Give at least one adaptation, extension or generalisation of the
         problem.

         Adaptation 1:
         A boy claims that when he left school X and joined school Y, he
         lowered the average IQ of both schools. Explain if this is
         possible.

         Adaptation 2:
         A striker in football is rated according to the average number of
         goals he scores in a game. Wayne had a higher average than

Carlos for all games over two years 2007 and 2008. Can we say that Wayne had a higher average than Carlos for at least one of the years 2007 and 2008?

Generalisation:
A striker in football is rated according to the average number of goals he scores in a game. Wayne had a higher average than Carlos for games in each of the years 2000 to 2010. Can we say that Wayne is a better striker than Carlos over the years 2000-2010?

Solution(s)

Solution 1
(a) If the IQ of the boy is less than the initial average IQ of school X, then when he leaves X, the average IQ of X will increase. Now, if the IQ of the boy is more than the initial average IQ of school Y, then when he joins Y, the average IQ of Y will also increase. Thus, the situation is possible.
(b) The following is a counterexample.
In 2007, Wayne scored 11 goals in 20 games, Carlos scored 1 goal in 2 games.
In 2008, Wayne scored 1 goal in 1 game, Carlos scored 18 goal in 19 games.
Over two years, Wayne scored 12 goals in 21 games, Carlos scored 19 goals in 21 games. Thus, Wayne has a better average in each year but a worse average over two years.

Solution 2
(a) Let IQ of boy = q, initial average IQ of school X = x, final average IQ of school X = x', initial number of students in school X = m, initial average IQ of school Y = y, final average IQ of school Y = y', and initial number of students in school Y = n.

Then $x' = \dfrac{mx - q}{m - 1} = x + \dfrac{x - q}{m - 1}$ and $y' = \dfrac{ny + q}{n + 1} = y + \dfrac{q - y}{n + 1}$.

If $y < q < x$, then $x' > x$ and $y' > y$, which is what we want.

Possible student responses

Students find it difficult to think of suitable values – guide them to understand that Guess-and-Check involves 'intelligent' guessing and adjusting the next guess from what has failed in the earlier guesses.

Assessment notes

Any numerical example for (a) and counterexample for (b) is acceptable.

Problem 14: Intersection of Two Squares

Heuristics to highlight

Draw a diagram; Consider a simpler problem (special case – tighten conditions); Relate to a similar problem

---

Two squares, each s on a side, are placed such that the corner of one square lies on the centre of the other. Describe, in terms of s, the range of possible areas representing the intersections of the two squares.
(from Schoenfeld, 1985)

Suitable hints for three of Pólya's stages

I   Understand the problem
(c)   Write down the heuristics you used to understand the problem.

Draw a diagram – Ensure that all the conditions are fulfilled: corner of one square lies on the centre of the other, squares are of equal size.

II   Devise a plan
(a)   Write down the key concepts that might be involved in solving the problem.

Areas of polygons.

(c)   Write out each plan concisely and clearly.

Plan 1   *Consider a simpler problem (special case – tighten conditions) – choose one or two special orientations of the squares for which the intersecting area can be easily calculated.*

Plan 2   *Relate to a similar problem - see how the area of one of the special cases transforms to the area of the general case.*

IV   Check and Expand

(a)   Write down how you checked your solution.

By using a dynamic geometry software.

(b)   Write down a sketch of any alternative solution(s) that you can think of.

Since this is a geometry problem involving only straight lines, one could think of using coordinate geometry to solve this problem. However, the formulation of this problem using coordinate geometry might be "messy" and needs careful consideration.

(c)   Give at least one adaptation, extension or generalisation of the problem.

Adaptation:
Two hexagons, each s on a side, are placed such that the corner of one hexagon lies on the centre of the other. Describe, in terms of s, the range of possible areas representing the intersections of the two hexagons.

Generalisation:
Two 2n-gons, each s on a side, are placed such that the corner of one 2n-gon lies on the centre of the other. Describe, in terms of s, the range of possible areas representing the intersections of the two 2n-gons.

Extension:
Two cubes, each s on a side, are placed such that the corner of one cube touches the centre of the other. Describe, in terms of s,

the range of possible volumes representing the intersections of the two cubes.

Solution(s)

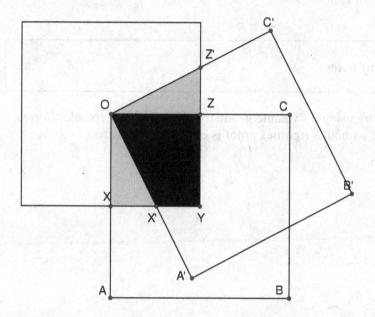

Consider the first position of the 'moving' square at OABC. It is easy to see that the area of the intersection = area of OXYZ = $\dfrac{s^2}{4}$.

At the second position OA'B'C', the intersection has lost the lightly shaded area OXX' but gained the lightly shaded area OZZ'. It is easy to show that the triangles OXX' and OZZ' are congruent and so the area of the intersection (now OX'YZ') has not changed. Thus, area of the intersection = $\dfrac{s^2}{4}$.

Possible student responses

Some students might only end up with considering the special case, without rigorous justification of why the areas should be equal in the other general cases – emphasise that 'Pattern is NOT proof' and that a general proof is needed.

Assessment notes

Checking by using a dynamic geometry software is acceptable. Correct conjecture without a rigorous proof is only partially correct.

Problem 15: Regions in a Circle

Heuristics to highlight
Draw a diagram; Look for patterns.

A circle has *n* points on its circumference. All possible chords are drawn and no three chords are concurrent. Find the number of regions that the chords divide the circle into for $n = 1, 2, 3, 4, 5, 6$.

Suitable hints for three of Pólya's stages

I   Understand the problem
(c)   Write down the heuristics you used to understand the problem.

Draw a diagram – Make sure that no three chords are concurrent.

II   Devise a plan
(a)   Write down the key concepts that might be involved in solving the problem.

Counting techniques.

(c)   Write out each plan concisely and clearly.

Draw a diagram, Look for patterns – Ensure that the diagram is large and neat, remember that 'Pattern is NOT proof' and one must understand what 'drives' the pattern.

IV   Check and Expand
(a)   Write down how you checked your solution.

Check how the 'new' regions are generated from older ones for n = 2, 3, 4.

(b)   Write down a sketch of any alternative solution(s) that you can think of.

Prove that given the number of regions for n points is S_n, then by adding the (n+1)-th point, drawing one chord after another and counting the additional regions formed, S_{n+1} can be obtained from S_n. The resulting recursive equation can be used to obtain the answers starting from $S_1 = 1$.

(c)   Give at least one adaptation, extension or generalisation of the problem.

Adaptation:
A circle has n points on its circumference. All possible chords are drawn and no three chords are concurrent. Find the number of triangles formed for n = 1, 2, 3, 4, 5, 6.

> *Generalisation:*
> *A circle has n points on its circumference. All possible chords are drawn and no three chords are concurrent. Find the number of regions that the chords divide the circle into.*

Solution(s)

Solution 1
By careful counting, we have the following numbers of regions: 1, 2, 4, 8, 16, 31.

Solution 2
Let the number of regions for n points, labeled in order as 1, 2, ..., n, be S_n. Clearly, $S_1 = 1$. Add a new point 0 between 1 and n.

Suppose $n \geq 2$ is even. Draw the chord 0-1. It does not cut any other chord so only 1 new region is created. Draw the next chord 0-2. It will cut all chords 1-x, $x \geq 3$. $(n-1)$ new regions will be formed. Draw the next chord 0-3. It will cut all chords 1-x and 2-x, $x \geq 4$. $2(n-3) + 1$ new regions will be formed. In general, the 0-j chord, where $1 \leq j \leq \frac{n}{2}$ will cut all chords i-x, where $i < j < x \leq n$, forming $(j-1)(n-j) + 1$ new regions. Thus, the number of new regions formed by chords 0-j, where $1 \leq j \leq \frac{n}{2}$ is $1 + (n-1) + (2(n-3) + 1) + \ldots + ((\frac{n}{2}-1)(\frac{n}{2}) + 1)$. By symmetry, the chords from 0-$(\frac{n}{2}+1)$ to 0-n will form the same number of new regions. Thus, $S_{n+1} = S_n + 2(1 + (n-1) + (2(n-3) + 1) + \ldots + ((\frac{n}{2}-1)(\frac{n}{2}) + 1)$.

Suppose $n \geq 1$ is odd. As before, the 0-1 chord will create one new region and in general, the 0-j chord, where $2 \leq j \leq \frac{n-1}{2}$ will cut all chords i-x, where $i < j < x \leq n$, forming $(j-1)(n-j) + 1$ new regions will be formed. Thus, the number of new regions formed by chords 0-j, where $1 \leq j \leq \frac{n-1}{2}$ is $1 + (n-1) + (2(n-3) + 1) + \ldots + ((\frac{n-1}{2}-1)(\frac{n+1}{2}) + 1)$. By symmetry, the chords from 0-$(\frac{n+3}{2})$ to 0-n will form the same number of

new regions. The chord $0\text{-}\frac{n+1}{2}$ will create $(\frac{n-1}{2})^2 + 1$. Thus, $S_{n+1} = S_n +$
$2(1 + (n{-}1) + 2(n{-}2) + \ldots + ((\frac{n-1}{2}{-}1)(\frac{n+1}{2}) + 1)) + (\frac{n-1}{2})^2 + 1$.

The answers can be obtained by recursion. Thus, we have:
$S_1 = 1, S_2 = 2$.
$S_3 = 2 + 2(1) = 4$.
$S_4 = 4 + 2(1) + (1)(2) = 4 + 4 = 8$.
$S_5 = 8 + 2(1 + 3) = 8 + 8 = 16$.
$S_6 = 16 + 2(1 + 4) + 2^2 + 1 = 16 + 15 = 31$.

Note: A closed form can be obtained by double summation.

Possible student responses

Students wrongly infer that $S_6 = 32$ from the earlier results – this is a fine opportunity to show that not only 'Pattern is NOT proof' but also that patterns can be misleading.

Assessment notes

A suitably labeled diagram showing 31 regions for $n = 6$ is acceptable.

Problem 16: Nice Numbers I

Heuristics to highlight

Solve a simpler problem (smaller numbers); Look for patterns; Use equations/algebra

---

A 'nice' number is a number that can be expressed as the sum of a string of two or more consecutive positive integers. Determine which of the numbers from 50 to 70 inclusive are nice.

(from Derek Holton)

Suitable hints for three of Pólya's stages

I   Understand the problem
(c)   Write down the heuristics you used to understand the problem.

Solve a simpler problem (smaller numbers) – Work out if 1, 2, ..., 8 are nice.

II   Devise a plan
(a)   Write down the key concepts that might be involved in solving the problem.

Factorisation.

(c)   Write out each plan concisely and clearly.

Plan 1   *Look for patterns – work out all of the numbers in the
                    required range of 50 to 70 and look out for patterns, for
                    example odd numbers behave in a particular way.*

Plan 2   *Solve a simpler problem (smaller numbers), Look for
                    patterns – work for the smaller numbers, say 1 to 20 to
                    find out which numbers don't seem nice, look out for
                    patterns. Prove by brute force for any not nice number in
                    the required range.*

IV   Check and Expand
(a)   Write down how you checked your solution.

*By showing that 32 is not nice using brute force. This follows
from our conjecture that all powers of 2 are not nice.*

(b)   Write down a sketch of any alternative solution(s) that you can
         think of.

*Use equations/algebra – work on the sum of consecutive integers
algebraically.*

(c)   Give at least one adaptation, extension or generalisation of the
         problem.

Adaptation:
*Determine which of the numbers from 100 to 130 inclusive
are nice.*

Generalisation:
Prove that an integer is nice if and only if it is not a power of 2.

Extension 2:
*A 'b-nice' number, where b is a positive integer, is a number that
can be expressed as the sum of a string of two or more
consecutive positive integers in at least b ways. Determine for*

each of the numbers from 50 to 70 inclusive, the maximum value of b for which they are b-nice.

Solution(s)

Solution 1

All odd integers n can be expressed as $\frac{n-1}{2} + \frac{n+1}{2}$. This takes care of 51, 53, ..., 69.

If n is a multiple of 3, it can be expressed as $(\frac{n}{3} - 1) + \frac{n}{3} + (\frac{n}{3} + 1)$. This takes care of 54, 60, 66.

If n is a multiple of 5, it can be expressed as $(\frac{n}{5} - 2) + (\frac{n}{5} - 1) + \frac{n}{5} + (\frac{n}{5} + 1) + (\frac{n}{5} + 2)$. This takes care of 50, 70.

$52 = 3 + 4 + 5 + 6 + 7 + 8 + 9 + 10$

$56 = 5 + 6 + 7 + 8 + 9 + 10 + 11$

$58 = 13 + 14 + 15 + 16$

$62 = 14 + 15 + 16 + 17$

$68 = 5 + 6 + 7 + 8 + 9 + 10 + 11 + 12$

The following proves that 64 is not nice:

$1+2+3+4+5+6+7+8+9+10 = 55 < 64 < 66 = 1+2+3+4+5+6+7+8+9+10+11$;

$2+3+4+5+6+7+8+9+10 < 3+4+5+6+7+8+9+10+11 = 63 < 64 < 65 = 2+3+4+5+6+7+8+9+10+11$;

$4+5+6+7+8+9+10+11 = 60 < 64 < 72 = 4+5+6+7+8+9+10+11+12$;

$5+6+7+8+9+10+11 < 6+7+8+9+10+11+12 = 63 < 64 < 68 = 5+6+7+8+9+10+11+12$;

$7+8+9+10+11+12 < 8+9+10+11+12+13 = 63 < 64 < 70 = 7+8+9+10+11+12+13$;

$9+10+11+12+13 < 10+11+12+13+14 < 11+12+13+14+15 = 63 < 64 < 69 = 9+10+11+12+13+14$;

$12+13+14+15 < 13+14+15+16 < 14+15+16+17 = 62 < 64 < 70 = 12+13+14+15+16$;

$15+16+17 = 48 < 64 < 16+17+18 < 17+18+19 < 18+19+20 < 19+20+21 < 20+21+22 = 63 < 64 < 66 = 15+16+17+18$;

$21+22 < 22+23 < ... < 31+32 = 63 < 64 < 66 = 21+22+23$.

Solution 2

We want to prove algebraically that 64 is not nice.

Suppose 64 is nice. Then for some positive integers a and b, we have the following.

$$64 = a + (a+1) + ... + (a+b)$$
$$= \frac{(a+a+b)(b+1)}{2}$$
$$= \frac{(2a+b)(b+1)}{2}$$
$$128 = (2a+b)(b+1).$$

It is clear that 128 does not have any odd factors. However, if b is even, then $b+1$ is odd, and if b is odd, then $2a+b$ is odd. This is a contradiction and thus 64 is not nice.

Possible student responses

Students give up, thinking that there are too many numbers to work on – suggest systematic work and advise that "Listing is like a train, it starts slow but picks up speed eventually."

Assessment notes

Failure to show rigorously that 64 is not nice will be graded as partially correct at best.

Problem 17: Nice Numbers II

Heuristics to highlight

Think of a related problem; Divide into cases; Use equations/algebra.

---

n is a positive integer that is not a power of 2. Show that *n* is a nice number.

(from Derek Holton)

Suitable hints for three of Pólya's stages

I   Understand the problem

(c)   Write down the heuristics you used to understand the problem.

Think of a related problem – recall how the odd numbers could be split by dividing by 2 and then taking the integers either side of the result.

II   Devise a plan

(a)   Write down the key concepts that might be involved in solving the problem.

Factorisation.

(c)   Write out each plan concisely and clearly.

Plan 1   *Divide into cases – work for multiples of 3, 5, 7, …*

Plan 2   *Divide into cases, use equations/algebra – consider whether to divide by a power of 2 or by an odd factor.*

IV   Check and Expand

(a)   Write down how you checked your solution.

Use algorithm for some numbers – 36, 47, 128×3.

(b)   Write down a sketch of any alternative solution(s) that you can think of.

(c)   Give at least one adaptation, extension or generalisation of the problem.

Adaptation:
Express 105 as the sum of the maximum number of consecutive integers.

Generalisation:
$n = p_1p_2...p_k$, *where the p_i-s are distinct odd primes. What is the maximum number of consecutive integers for which n can be expressed as a sum?*

Extension:
A 'b-nice' number, where b is a positive integer, is a number that can be expressed as the sum of a string of two or more consecutive positive integers in at least b ways. $n = pq$, where p and q are odd primes. Show that n is a 2-nice number.

Solution(s)

Let $n = 2^p q$, where p is a nonnegative integer and q is odd.
If $q < 2^{p+1}$, then $n = (2^p - \frac{q-1}{2}) + ... + 2^p + ... + (2^p + \frac{q-1}{2})$.
If $q > 2^{p+1}$, then $n = (\frac{q-1}{2} - 2^p + 1) + ... + \frac{q-1}{2} + \frac{q+1}{2} + ... + (\frac{q+1}{2} + 2^p - 1)$.

In words:
If $q < 2^{p+1}$, then divide n by q (obtaining 2^p), and take $\frac{q-1}{2}$ integers before and after 2^p.
If $q > 2^{p+1}$, then divide n by 2^{p+1} (obtaining $\frac{q}{2}$), take 2^p integers before and including $\frac{q-1}{2}$, and take 2^p integers after and including $\frac{q+1}{2}$.

Possible student responses

Students ask what is required of them to prove – explain that stating an algorithm (procedure stated mathematically) that produces a sum of consecutive integers is required.

Assessment notes

A clearly stated word explanation of the algorithm is acceptable.

Appendix I

The Mathematics Practical Worksheet

Problem

> Print the problem here.

Instructions

1. You may proceed to complete the worksheet doing stages I – IV.
2. If you wish, you have 15 minutes to solve the problem without explicitly using Polya's model. Do your work in the space for Stage III.
   - If you are stuck after 15 minutes, use Polya's model and complete all the stages I – IV.
   - If you can solve the problem, you must proceed to do stage IV – Check and Expand.

I   Understand the problem

(You may have to return to this section a few times. Number each attempt to understand the problem accordingly as Attempt 1, Attempt 2, etc.)

(a) Write down your feelings about the problem. Does it bore you? scare you? challenge you?

(b) Write down the parts you do not understand now or that you misunderstood in your previous attempt.

(c) Write down your attempt to understand the problem; and state the heuristics you used.

<u>Attempt 1</u>

II Devise a plan

(You may have to return to this section a few times. Number each new plan accordingly as Plan 1, Plan 2, etc.)

(a) Write down the key concepts that might be involved in solving the problem.

(b) Do you think you have the required resources to implement the plan?

(c) Write out each plan concisely and clearly.

<u>Plan 1</u>

III   Carry out the plan

(You may have to return to this section a few times. Number each implementation accordingly as Plan 1, Plan 2, etc., or even Plan 1.1, Plan 1.2, etc. if there are two or more attempts using Plan 1.)

(i) Write down in the *Control* column, the key points where you make a decision or observation, for eg., go back to check, try something else, look for resources, or totally abandon the plan.

(ii) Write out each implementation in detail under the *Detailed Mathematical Steps* column.

Detailed Mathematical Steps	*Control*
<u>Attempt 1</u>	

IV   **Check and Expand**

(a) Write down how you checked your solution.

(b) Write down your level of satisfaction with your solution. Write down a sketch of any alternative solution(s) that you can think of.

(c) Give one or two adaptations, extensions or generalisations of the problem. Explain succinctly whether your solution structure will work on them.

Appendix II

Sample of Student's Work

Practical Worksheet

Problem

The Lockers Problem
The new school has exactly 343 lockers numbered 1 to 343, and exactly 343 students. On the first day of school, the students meet outside the building and agree on the following plan. The first student will enter the school and open all the lockers. The second student will then enter the school and close every locker with an even number. The third student will then 'reverse' every third locker; i.e. if the locker is closed, he will open it, and if the locker is open, he will close it. The fourth student will reverse every fourth locker, and so on until all 343 students in turn have entered the building and reversed the relevant lockers. Which lockers will finally remain open?

Instructions

- You may proceed to complete the worksheet doing stages I – IV.
- If you wish, you have 15 minutes to solve the problem without explicitly using Polya's model. Do your work in the space for Stage III.
  - If you are stuck after 15 minutes, use Polya's model and complete all the stages I – IV.
  - If you can solve the problem, you must proceed to do stage IV – Check and Extend.

I   Understand the problem

(You may have to return to this section a few times. Number each attempt to understand the problem accordingly as Attempt 1, Attempt 2, etc.)
(a) Write down your feelings about the problem. Does it bore you? scare you? challenge you?
(b) Write down the parts you do not understand now or that you misunderstood in your previous attempt.
(c) Write down your attempt to understand the problem; and state the heuristics you used.

Attempt 1

(a) The question has sparked my interest due to the unusual nature of the problem. I believe that this problem would prove to be challenging, and conventional methods taught in school may not be sufficient to tackle this problem.

(b) N/A

(c) Simulation of 10 lockers, 10 students.

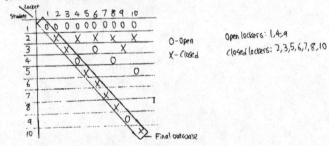

Open lockers: 1, 4, 9
Closed lockers: 2, 3, 5, 6, 7, 8, 10

II   **Devise a plan**
(You may have to return to this section a few times. Number each new plan accordingly as Plan 1, Plan 2, etc.)
(a) Write down the key concepts that might be involved in solving the problem.
(b) Do you think you have the required resources to implement the plan?
(c) Write out each plan concisely and clearly.

<u>Plan 1</u>

(a) Finding number of factors a number has

(b) Yes

1. Consider any integer n, where $1 \leq n \leq 343$.

2. Label the students from 1 to 343, according to the lockers they reverse.

3. Observe which students open locker n.

4. Deduce a relation from n, and prove if possible.

5. Identify conditions for n to be open/closed.

6. If successful, find all open lockers.

2

III   Carry out the plan

(You may have to return to this section a few times. Number each implementation accordingly as Plan 1, Plan 2, etc., or even Plan 1.1, Plan 1.2, etc. if there are two or more attempts using Plan 1.)

(i)   Write down in the *Control* column, the key points where you make a decision or observation, for e.g., go back to check, try something else, look for resources, or totally abandon the plan.

(ii) Write out each implementation in detail under the *Detailed Mathematical Steps* column.

Detailed Mathematical Steps	Control
Attempt 1	
Let student n be the student that reverses every nth locker.	Definining variables
Consider locker 20.	
By trial and error,	
Student #1 opens the locker,	
#2 closes	Attempt to find a relation between locker number and student number
#4 opens	
#5 closes	
#10 opens	
#20 closes	
∴ Deduction: Student m only reverses locker ℓ if $m \mid \ell$.	Make a claim
This is in fact true, since any student m only reverses locker numbers $m, 2m, 3m, \ldots$, which are multiples of m. This means that if ℓ is a multiple of m, then the student will reverse it. Thus, a locker ℓ is reversed by students whose number divides ℓ.	Justifying the claim
Observation: If a locker number ℓ has odd number of factors, then it will ultimately remain open.	Make another claim
If ℓ has an odd number of factors, then an odd number of students reversed the locker. If the locker starts out closed, it _must_ end up open.	Justifying the claim

3

It remains to find locker numbers which have an odd number of factors.

Suppose the locker number ℓ has the following prime factorisation:

$$\ell = p_1^{q_1} \, p_2^{q_2} \dots p_y^{q_y}$$

Then it has the following factor:

$$p_1^{r_1} p_2^{r_2} \dots p_y^{r_y} \mid p_1^{q_1} p_2^{q_2} \dots p_y^{q_y}, \quad 0 \le r_i \le q_i, \quad 1 \le i \le y$$

r_i can take $q_i + 1$ values, and by simple combinatorics, the total number of factors ℓ can have is given by

$$\prod_{i=1}^{y} (q_i + 1) = \text{number of factors of } \ell. \qquad (*)$$

Since we want number of factors to be odd, there must be no even number in the multiplication. That is,

$$(q_1 + 1), (q_2 + 1), \dots, (q_y + 1) \text{ are all odd numbers.}$$

Equivalently,

$$q_1, q_2, q_3, \dots, q_y \text{ are all even numbers.}$$

This means that $q_1, q_2, \dots, q_y$ will yield an integer if divided by 2. Also, this relates to $\sqrt{\ell}$ having an integer value $\Rightarrow \ell$ is a square number.

Conclusion: All lockers whose number is a perfect square will be open.

If a locker whose number ℓ' is not a square number, then there must exist in the prime factorisation a $p_x^{q_x}$ where q_x is odd. This means that $(q_x + 1)$ is even, thus ℓ' has an even number of factors by $(*)$. This relates to an even number of students reversing the locker, which means if it starts out closed, it will ultimately be closed.

4

$\therefore$ lockers that remain open are numbers under 343 and are perfect squares. Namely: 1, 4, 9, 16, 25, 36, 49, 64, 81, 100, 121, 144, 169, 196, 225, 256, 289, 324

(right margin annotations)

Obtaining the necessary condition for locker ℓ to remain open

Showing that the condition is sufficient

IV   **Check and Expand**

    (a) Write down how you checked your solution.

    (b) Write down your level of satisfaction with your solution. Write down a sketch of any alternative solution(s) that you can think of.

    (c) Give one or two adaptations, extensions or generalisations of the problem. Explain succinctly whether your solution structure will work on them.

(a) From the simple 10-student/locker simulation, the result is correct. Square numbers are checked to be the only integers that have an odd number of factors, and are thus the only lockers that are open.

(b) Satisfied.

<u>Alternative solution</u>

It is noted that if student m opens locker ℓ, then student $\frac{\ell}{m}$ opens locker ℓ as well.

This is generally true, since both m and $\frac{\ell}{m}$ are integers less than ℓ, and both are factors of ℓ.

Thus, if $a_1, a_2, a_3, \ldots, a_n$ are factors of ℓ less than $\sqrt{\ell}$,

a_1 opens $\rightarrow \frac{\ell}{a_1}$ closes

a_2 opens $\rightarrow \frac{\ell}{a_2}$ closes

$\vdots$

a_n opens $\rightarrow \frac{\ell}{a_n}$ closes

there does not exist any a_i which has the same as any $\frac{\ell}{a_j}$, since all $a_i \leq \sqrt{n}$, and $a_j \gtrsim \sqrt{n}$.

<u>unless</u> $a_i = \frac{\ell}{a_i} = \sqrt{\ell}$. In that case a_i will open, and no $\frac{\ell}{a_i}$ will close, ultimately resulting in the locker being open. For such a_i to exist, $\sqrt{\ell}$ must be an integer.

$\therefore \ell$ must be a perfect square.

Ans: $1, 4, 9, 16, 25, 36, 49, 64, 81, 100, 121, 144, 169, 196, 225, 256, 289, 324$

	Date	No.

Problem | Extension

Consider the same locker's problem, except that the kth student now opens every k^{th} locker. If there are 343 lockers and 343 students, Which lockers remain open?

Solution | Consider locker ℓ with the following prime factorisation.
$$\ell = p_1^{q_1} p_2^{q_2} \dots p_n^{q_n}$$

Similar to the original case, we want an odd number of students to reverse the locker, so that it ultimately ends up open. Also, if ℓ is a multiple of n^2, then student n will reverse the locker. Thus, if ℓ has an odd number of perfect square factors, then it will be open.

The problem is thus reduced to finding number of perfect square factors of ℓ. ℓ has the following perfect Square factor:
$$p_1^{2r_1} p_2^{2r_2} p_3^{2r_3} \dots p_n^{2r_n} \mid p_1^{q_1} p_2^{q_2} \dots p_n^{q_n}$$
where $0 \le 2r_i \le q_i$, $1 \le i \le n$.

Thus, r_i can take $\lfloor \frac{q_i}{2} \rfloor + 1$ values, and by simple combinatorics, the number of perfect square factors of ℓ is given by
$$\prod_{i=1}^{n} (\lfloor \frac{q_i}{2} \rfloor + 1) = \text{number of square factors of } \ell$$

As similar to previously, all $(\lfloor \frac{q_i}{2} \rfloor + 1)$ must be odd, and thus all $\lfloor \frac{q_i}{2} \rfloor$ must be even. That is,
$$\lfloor \frac{q_i}{2} \rfloor \equiv 0 \bmod 2$$

Solving for q_i:
$$q_i \equiv 0, 1 \bmod 4$$

Thus, all powers of p must be in the form $(4n+1)$ or $(4n)$.

∴ All lockers whose powers is congruent to 0 or 1 modulo 4 will have the lockers left open.
In prime factorisation

★ This may be used to prove that if a student k opens every k^{th} th locker, then
$$\prod_{i=1}^{n} (\lfloor \frac{q_i}{k} \rfloor + 1)$$

has to be odd for ℓ to be open. The results, however, may not be appreciated.

Appendix III

Rubrics for Assessment

Name: _____

Polya's Stages		
	Descriptors/Criteria *(evidence suggested/indicated on practical sheet or observed by teacher)*	**Marks Awarded**
Correct Solution		
Level 3	Evidence of complete use of Pólya's stages – UP + DP + CP; and when necessary, appropriate loops. [10 marks]	
Level 2	Evidence of trying to understand the problem and having a clear plan – UP + DP + CP. [9 marks]	
Level 1	No evidence of attempt to use Pólya's stages. [8 marks]	
Partially Correct Solution *(solve significant part of the problem or lacking rigour)*		
Level 3	Evidence of complete use of Pólya's stages – UP + DP + CP; and when necessary, appropriate loops. [8 marks]	
Level 2	Evidence of trying to understand the problem and having a clear plan – UP + DP + CP. [7 marks]	
Level 1	No evidence of attempt to use Pólya's stages. [6 marks]	

Incorrect Solution		
Level 3	Evidence of complete use of Polya's stages – UP + DP + CP; and when necessary, appropriate loops. [6 marks]	
Level 2	Evidence of trying to understand the problem and having a clear plan – UP + DP + CP. [5 marks]	
Level 1	No evidence of attempt to use Pólya's stages. [0 marks]	

Heurisitcs		
	Descriptors/Criteria *(evidence suggested/indicated on practical sheet or observed by teacher)*	**Marks Awarded**
Correct Solution		
Level 2	Evidence of appropriate use of heuristics. [4 marks]	
Level 1	No evidence of heuristics used. [3 marks]	
Partially Correct Solution *(solve significant part of the problem or lacking rigour)*		
Level 2	Evidence of appropriate use of heuristics. [3 marks]	
Level 1	No evidence of heuristics used. [2 marks]	
Incorrect Solution		
Level 2	Evidence of appropriate use of heuristics. [2 marks]	

Level 1	No evidence of heuristics used. [0 marks]	
Checking and Expanding		
	Descriptors/Criteria *(evidence suggested/indicated on practical sheet or observed by teacher)*	**Marks Awarded**
Checking		
Level 2	Checking done – mistakes identified and correction attempted by cycling back to UP, DP, or CP, until solution is reached. [1 mark]	
Level 1	No checking, or solution contains errors. [0 marks]	
Alternative Solutions		
Level 3	Two or more correct alternative solutions. [2 marks]	
Level 2	One correct alternative solution. [1 mark]	
Level 1	No alternative solution. [0 marks]	
Extending, Adapting & Generalizing		
Level 4	More than one related problem with suggestions of correct solution methods/strategies; or one **significant** related problem, with suggestion of correct solution method/strategy; or one **significant** related problem, with explanation why method of solution for original problem cannot be used. [3 marks]	
Level 3	One related problem with suggestion of correct solution method/strategy. [2 marks]	

Level 2	One related problem given but without suggestion of correct solution method/strategy. [1 mark]	
Level 1	None provided [0 marks]	

Hints given:

Marks deducted: _____

Total marks: _____